GEOTECHNICAL SPECIAL PUBLICATION NO. 115

EXPANSIVE CLAY SOILS AND VEGETATIVE INFLUENCE ON SHALLOW FOUNDATIONS

PROCEEDINGS OF GEO-INSTITUTE SHALLOW FOUNDATION AND SOIL PROPERTIES COMMITTEE SESSIONS AT THE ASCE 2001 CIVIL ENGINEERING CONFERENCE

SPONSORED BY
The Geo-Institute of the American Society of Civil Engineers

October 10–13, 2001
Houston, Texas

EDITED BY
C. Vipulanandan
Marshall B. Addison
Michael Hasen

1801 ALEXANDER BELL DRIVE
RESTON, VIRGINIA 20191–4400

Abstract: This proceedings *Expansive Clay Soils and Vegetative Influence on Shallow Foundations*, contains papers on the effects of trees and moisture movements on various lightly loaded foundations on expansive clays. Papers included in the proceedings cover such topics as tree root damage to buildings, predicting volume changes in expansive soils due to environmental factors including vegetation, drought, expansive clay soil problems and solutions throughout the world, soil-structure interaction, field test program to obtain geotechnical properties, depth of wetting and active zone in expansive soils, slabs on expansive clay with vegetation considerations, volume change coefficient of expansive clay, in-situ modification of active clays for shallow foundation remediation, cracking of pavement due to drying shrinkage, soil suction measurements and data base on under-slab moisture contents in Texas.

Library of Congress Cataloging-in-Publication Data

Civil Engineering Conference (2001 : Houston, Tex.)
Expansive clay soils and vegetative influence on shallow foundations : proceedings of Geo-Institute Shallow Foundation and Soil Properties Committee Sessions at the ASCE 2001 Civil Engineering Conference / sponsored by the Geo-Institute of the American Society of Civil Engineers, October 10-13, 2001, Houston, Texas ; edited by C. Vipulanandan, Marshall B. Addison, Michael Hasen.
p. cm.— (Geotechnical special publication ; no. 115)
Includes bibliographical references and index.
ISBN 0-7844-0592-1
1. Foundations—Congresses. 2. Soil mechanics—Congresses. I. Vipulanandan, Cumaraswamy, 1956- . II. Addison, Marshall B. III. Hasen, Michael. IV. American Society of Civil Engineers. Geo-Institute. V. Title. VI. Series.

TA775.C58 2001
624.1'5136--dc21 2001053327

Library of Congress Catalog Card No: 2001053327
ISBN 0-7844-0592-1
Manufactured in the United States of America.

Geotechnical Special Publications

1 *Terzaghi Lectures*
2 *Geotechnical Aspects of Stiff and Hard Clays*
3 *Landslide Dams: Processes, Risk, and Mitigation*
4 *Tiebacks for Bulkheads*
5 *Settlement of Shallow Foundation on Cohesionless Soils: Design and Performance*
6 *Use of In Situ Tests in Geotechnical Engineering*
7 *Timber Bulkheads*
8 *Foundations for Transmission Line Towers*
9 *Foundations & Excavations in Decomposed Rock of the Piedmont Province*
10 *Engineering Aspects of Soil Erosion, Dispersive Clays and Loess*
11 *Dynamic Response of Pile Foundations-Experiment, Analysis and Observation*
12 *Soil Improvement: A Ten Year Update*
13 *Geotechnical Practice for Solid Waste Disposal '87*
14 *Geotechnical Aspects of Karst Terrains*
15 *Measured Performance Shallow Foundations*
16 *Special Topics in Foundations*
17 *Soil Properties Evaluation from Centrifugal Models*
18 *Geosynthetics for Soil Improvement*
19 *Mine Induced Subsidence: Effects on Engineered Structures*
20 *Earthquake Engineering & Soil Dynamics II*
21 *Hydraulic Fill Structures*
22 *Foundation Engineering*
23 *Predicted and Observed Axial Behavior of Piles*
24 *Resilient Moduli of Soils: Laboratory Conditions*
25 *Design and Performance of Earth Retaining Structures*
26 *Waste Containment Systems: Construction, Regulation, and Performance*
27 *Geotechnical Engineering Congress*
28 *Detection of and Construction at the Soil/Rock Interface*
29 *Recent Advances in Instrumentation, Data Acquisition and Testing in Soil Dynamics*
30 *Grouting, Soil Improvement and Geosynthetics*
31 *Stability and Performance of Slopes and Embankments II*
32 *Embankment of Dams-James L. Sherard Contributions*
33 *Excavation and Support for the Urban Infrastructure*
34 *Piles Under Dynamic Loads*
35 *Geotechnical Practice in Dam Rehabilitation*
36 *Fly Ash for Soil Improvement*
37 *Advances in Site Characterization: Data Acquisition, Data Management and Data Interpretation*
38 *Design and Performance of Deep Foundations: Piles and Piers in Soil and Soft Rock*
39 *Unsaturated Soils*
40 *Vertical and Horizontal Deformations of Foundations and Embankments*
41 *Predicted and Measured Behavior of Five Spread Footings on Sand*
42 *Serviceability of Earth Retaining Structures*
43 *Fracture Mechanics Applied to Geotechnical Engineering*
44 *Ground Failures Under Seismic Conditions*
45 *In Situ Deep Soil Improvement*
46 *Geoenvironment 2000*
47 *Geo-Environmental Issues Facing the Americas*
48 *Soil Suction Applications in Geotechnical Engineering*
49 *Soil Improvement for Earthquake Hazard Mitigation*
50 *Foundation Upgrading and Repair for Infrastructure Improvement*
51 *Performance of Deep Foundations Under Seismic Loading*
52 *Landslides Under Static and Dynamic Conditions-Analysis, Monitoring, and Mitigation*
53 *Landfill Closures-Environmental Protection and Land Recovery*
54 *Earthquake Design and Performance of Solid Waste Landfills*
55 *Earthquake-Induced Movements and Seismic Remediation of Existing Foundations and Abutments*
56 *Static and Dynamic Properties of Gravelly Soils*
57 *Verification of Geotechnical Grouting*
58 *Uncertainty in the Geologic Environment*
59 *Engineered Contaminated Soils and Interaction of Soil Geomembranes*
60 *Analysis and Design of Retaining Structures Against Earthquakes*

61 *Measuring and Modeling Time Dependent Soil Behavior*
62 *Case Histories of Geophysics Applied to Civil Engineering and Public Policy*
63 *Design with Residual Materials: Geotechnical and Construction Considerations*
64 *Observation and Modeling in Numerical Analysis and Model Tests in Dynamic Soil-Structure Interaction Problems*
65 *Dredging and Management of Dredged Material*
66 *Grouting: Compaction, Remediation and Testing*
67 *Spatial Analysis in Soil Dynamics and Earthquake Engineering*
68 *Unsaturated Soil Engineering Practice*
69 *Ground Improvement, Ground Reinforcement, Ground Treatment: Developments 1987-1997*
70 *Seismic Analysis and Design for Soil-Pile-Structure Interactions*
71 *In Situ Remediation of the Geoenvironment*
72 *Degradation of Natural Building Stone*
73 *Innovative Design and Construction for Foundations and Substructures Subject to Freezing and Frost*
74 *Guidelines of Engineering Practice for Braced and Tied-Back Excavations*
75 *Geotechnical Earthquake Engineering and Soil Dynamics III*
76 *Geosynthetics in Foundation Reinforcement and Erosion Control Systems*
77 *Stability of Natural Slopes in the Coastal Plain*
78 *Filtration and Drainage in Geotechnical/Geoenvironmental Engineering*
79 *Recycled Materials in Geotechnical Applications*
80 *Grouts and Grouting: A Potpourri of Projects*
81 *Soil Improvement for Big Digs*
82 *Risk-Based Corrective Action and Brownfields Restorations*
83 *Design and Construction of Earth Retaining Systems*
84 *Effects of Construction on Structures*
85 *Application of Geotechnical Principles in Pavement Engineering*
86 *Big Digs Around the World*
87 *Jacked Tunnel Design and Construction*
88 *Analysis, Design, Construction, and Testing of Deep Foundations*
89 *Recent Advances in the Characterization of Transportation Geo-Materials*
90 *Geo-Engineering for Underground Facilities*
91 *Special Geotechnical Testing: Central Artery/Tunnel Project in Boston, Massachusetts*
92 *Behavioral Characteristics of Residual Soils*
93 *National Geotechnical Experimentation Sites*
94 *Performance Confirmation of Constructed Geotechnical Facilities*
95 *Soil-Cement and Other Construction Practices in Geotechnical Engineering*
96 *Numerical Methods in Geotechnical Engineering: Recent Developments*
97 *Innovations and Applications in Geotechnical Site Characterization*
98 *Pavement Subgrade, Unbound Materials, and Nondestructive Testing*
99 *Advances in Unsaturated Geotechnics*
100 *New Technological and Design Developments in Deep Foundations*
101 *Slope Stability 2000*
102 *Trends in Rock Mechanics*
103 *Advances in Transportation and Geoenvironmental Systems Using Geosynthetics*
104 *Advances in Grouting and Ground Modification*
105 *Environmental Geotechnics*
106 *Geotechnical Measurements: Lab & Field*
107 *Soil Dynamics and Liquefaction 2000*
108 *Use of Geophysical Methods in Construction*
109 *Educational Issues in Geotechnical Engineering*
110 *Computer Simulation of Earthquake Effects*
111 *Judgment and Innovation: The Heritage and Future of the Geotechnical Engineering Profession*
112 *Soft Ground Technology*
113 *Foundations and Ground Improvement*
114 *Soils Magic*
115 *Expansive Clay Soils and Vegetative Influence on Shallow Foundations*

Preface

It is estimated that damage to buildings, pavements, and other structures constructed at sites with expansive soils exceeds several billion dollars. Designing and constructing lightly loaded foundations on sites with expansive soils and trees is a challenge. Under these conditions the volume change in soil and moisture movement induced by tree roots are important factors to the long-term performance of the structures. This conference on vegetative effects and expansive clays on shallow foundations was conceived as a forum to review current state-of-the-practice in foundation design, maintenance, and remediation. One objective of this geotechnical mini conference and the resulting proceedings is to provide an update on the effects of tree roots and moisture movement on lightly loaded foundation damage.

This conference also provides an effective means of sharing recent technological advances, engineering applications, and research results among practitioners, researchers, and designers. Designing, maintaining, and remediating shallow foundations on expansive clay with various vegetation influence will be discussed during the conference.

This proceedings contains some of the papers that were presented at the ASCE Annual Convention held in Houston, Texas, October 10–14, 2001. These sessions were jointly sponsored by the Shallow Foundation and Soil Properties Committees of the Geo-Institute, ASCE. Papers published in this proceedings were peer reviewed for its content and quality. The standards for peer review were essentially the same as those for papers being reviewed for possible publication in the ASCE Journal of Geotechnical and Geoenvironmental Engineering (i.e., at least two reviewers). All papers are eligible for discussion in the ASCE Journal of Geotechnical and Geoenvironmental Engineering as well as for ASCE awards.

The Editors would like to thank the paper authors for their cooperation under a tight schedule. Reviewers for this publication were:

M.B. Addison
M. Hasen
S. Javed
R.G. McKeen
K. Meyer
K.K. Muraleetharan
J.B. Nevels
J.B Park
A.J. Puppala
M.M. Reda
L.G. Schwarz
D. Snethen
A. Stevens
V. Tandon
C. Vipulanandan

Reviewers' constructive and timely reviews are very much appreciated.

Editors
C.Vipulanandan (Vipu)
Director, Center for Innovative Grouting Materials and Technology (CIGMAT), University of Houston, Houston, Texas

Marshall B. Addison
Consulting Geotechnical Engineer, Arlington, Texas

Michael Hasen
Vice President, HVJ Associates, Houston, Texas

Contents

Tree Root Damage to Buildings

Giles Biddle[1]

Abstract

With clay soils and temperate climatic conditions, trees are the biggest cause of soil drying and subsidence, resulting in foundation movement and damage. Understanding how trees dry the soil, and how this drying can be controlled, is essential for decisions on methods of preventing or remedying this damage.

The UK experience

The problems of tree root damage to buildings involve sociological as well as technical engineering issues. The cause of damage may be simply the result of tree roots causing drying and shrinkage of a clay soil below foundations, but in the UK the perception of the property owner to this damage has changed radically in recent years.

Subsidence damage in the UK as a result of tree root activity has always been a regular occurrence, but was accepted as part of the normal performance of a building. Cracks might appear in the summer, particularly in a dry year, but were merely repaired as part of routine owner maintenance. All that started to change in 1971 when, at the request of mortgage lenders, insurance companies first offered subsidence coverage on domestic properties. Coverage was provided without alteration of premiums in anticipation that subsidence was only a minor problem. Initially there were few claims, but in 1976 there was a severe drought, at that time considered the worst for 250 years. Extensive damage occurred, and the public woke up to the fact that insurers were liable, and that their properties could be repaired and redecorated at the insurer's expense. Insurers hoped this was a one-off experience, but once the public had been made aware of subsidence, they continued their demands for repair. The number of claims and cost of repairs soon started to escalate, aided by dry spells in 1984 and 3 dry years in 1989/'90/'91, so that claims are now far above the 1976 levels, despite generally less extreme climatic conditions (Fig. 1). In dry years, subsidence is now the greatest cause for insurance payouts. This does not imply that the damage is getting worse, but simply

---

[1] Dr P.G. Biddle, OBE, MA, DPhil, FArborA. Principal, P.G. Biddle Arboricultural Consultants, Willowmead, Ickleton Road, Wantage, OX12 9JA, UK; biddle@willowmead.co.uk; phone 00 44 1235 762478; fax 00 44 1235 768034

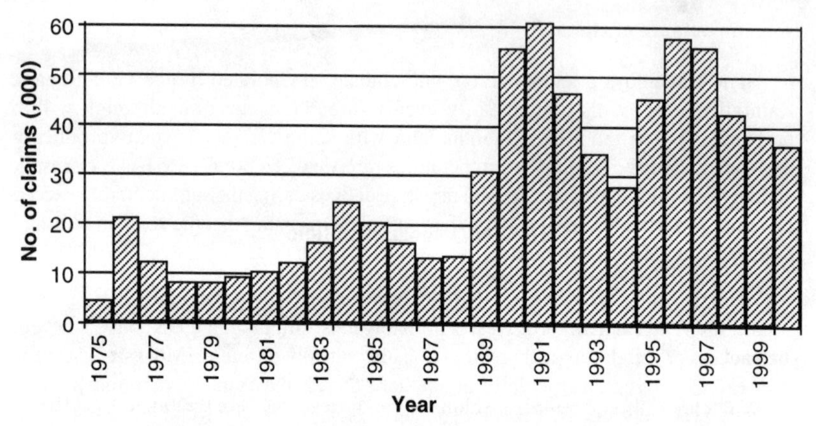

Figure 1. Number of insurance claims for subsidence since start of coverage.(Data by ABI)

that the public is far less tolerant of even very minor levels of damage and perceive such damage as a major cause for alarm.

The contract between insurers and their policyholders is now crucial to dealing with subsidence. In the UK, the Institution of Structural Engineers have recently published guidance (ISE, 2000) which defines a realistic threshold of damage, but there is little sign of this being adopted, and claims continue to be accepted for far lesser amounts of damage. This has far-reaching consequences for the many professionals who are involved. Engineers must ensure that even minor cracking is accurately diagnosed and effectively repaired. Arborists must try to identify and deal with trees or shrubs which either have, or may in the future, cause damage, whilst at the same time trying to ensure that unnecessary removal of trees does not damage the urban environment. Builders must ensure that new properties or extensions will not be damaged in the future. Surveyors, who advise a prospective purchaser, must check for any evidence of damage, and try to assess whether the property may be vulnerable in the future. All of these professionals act against a background of ever higher public expectations, and a fear of negligence claims. Although they provide the technical expertise, when it comes to dealing with the claim, it is loss adjusters, who often have little technical experience but act as the intermediary between insurers and their policyholder, who usually have the pivotal role in determining the type of investigations and the remedial action.

Mercifully the last few summers have been wetter than usual, which has helped to contain the problem, but there is little doubt that the next dry summer will produce a further major escalation. Insurers have the ability to prevent this from occurring by insisting on a more rational attitude by the public. If insurers were the only ones to suffer, they could be left to their fate, but sadly their decisions also have major environmental repercussions - trees are seen as the culprits and demands made for their removal, often irrespective of whether they pose any genuine threat.

The importance of climate

The UK enjoys a temperate oceanic climate, ameliorated by the Gulf Stream. Rainfall is usually distributed fairly evenly throughout the year although with a preponderance in the winter months, and with some weather systems producing prolonged periods of drought. Temperatures rarely exceed 30° C (86° F). As a result, under grass or similar vegetation, soil moisture deficits during the summer rarely exceed 150 mm, and winter rainfall is more than sufficient to ensure that the soil can return to field capacity during each winter.

The size of the USA ensures that different areas are subject to a far wider range of conditions. The Pacific North West is probably most similar to the UK, but elsewhere the patterns of soil drying and precipitation are radically different. Most significantly, over extensive areas, particularly in southern States, the annual evapotranspiration exceeds precipitation, so that the soil never returns to field capacity, but stays permanently desiccated.

This is an important distinction between the UK and USA and may mean that a different approach is required in the methods for investigating and remedying damage. A simple example of this different approach is how we consider our clay soils. In the UK we describe them as 'shrinkable', as the most significant movement which occurs is shrinkage as they dry during summer months. They do, of course, expand during the winter months as they return to field capacity. By contrast, over much of the USA (and in the title to these proceedings), clay soils are described as 'expansive', because of their propensity to rehydrate and expand if evapotranspiration is curtailed. This distinction is far more than a mere difference in terminology, as there are underlying fundamental differences in the patterns of soil drying which will occur under different climatic regimes.

The patterns of soil drying described in later sections of this paper are those applicable to UK conditions. However, where possible, consideration will be given to the differences which might occur under more arid conditions with expansive clay soils.

Factors affecting root growth

If we are to understand how root systems develop and the resulting distribution of roots and soil drying, one must consider the factors which influence root growth. These will be the same, irrespective of geographical location.

There is a tendency to think of the root system of a tree as a mirror image of the aerial parts. This would be grossly misleading. In general, a better analogy is to think of a tree as a wine glass, with the cup as the crown, the stem as the trunk, and the base as the root system. This helps to emphasise that 90% of the root system is usually within the surface 600 mm, with only a small proportion of roots extending to greater depth.

The structural roots which distribute the load of the trunk and anchor the tree to the soil are optimised to achieve uniform stress over their whole surface, and as a result are usually uniformly and radially distributed around the trunk. Their biomechanical efficiency in supporting and anchoring the tree should be the envy of any engineer.

Beyond the point where the tree is anchored, the root system rapidly subdivides into a mass of fine conducting roots which support the fine feeder roots. The growth of these fine feeder roots is opportunistic, proliferating where conditions are most conducive for obtaining the water and nutrients which are essential for the tree. The main factors which influence their distribution are the availability of water, an adequate supply of oxygen for respiration, other soil gases below their toxic level, suitable soil temperature and absence of mechanical impedance. If soil conditions are uniform, there would be a uniform distribution of roots around the trunk, but such uniformity is rarely present, particularly in the disturbed soils of the urban environment. As a result, the distribution of the conducting and fine feeder root is usually very irregular, reflecting any small variations in soil conditions.

Water uptake is restricted to the fine feeder roots. In appearance they are white, and heavily branched to maximise contact with the soil, and often with a zone of unicellular outgrowths, known as root hairs, which further increase the surface area. The fine white roots are short-lived, usually dying after a few weeks, although a proportion will continue growth, turning brown as they become lignified and suberised, to form conducting roots.

Apart from the roots of some species which have developed special means of coping with anaerobic conditions, all of these roots must have oxygen for their respiration. In soils of low permeability, such as some heavy clays, rates of gaseous diffusion can limit the depth of root penetration. The axial pressure exerted by the root tip is limited, so that mechanical impedance is also a major constraint in some soils, particularly clays which have a fine pore size and low porosity. Pore size and porosity of the soil are functions of the bulk density. In general terms, root growth is likely to be significantly reduced in soil with a bulk density in excess of 1.2 g/cm3, and to cease above 1.8 g/cm3. Penetrometers are used to measure the resistance of the soil to root growth, but different soils can exert a very different resistance and produce a different relationship between penetrometer readings and growth. In non-cohesive soils, the constant rate penetrometer is recommended, as this instrument mimics root behaviour in that it measures the pressure required to push through the soil. In non cohesive soils the rate of root elongation will diminish at penetrometer resistance in excess of a value of between 0.8 to 1.2 MPa, with little or no growth occurring above 4 MPa. In clay soils growth will be restricted at substantially lower values than these, but few experiments have been made on these soils.

In practice, many clay soils have a bulk density in excess of 1.8 g/cm^2. Furthermore, their penetrometer resistance increases markedly as the soil dries with readings in excess of 6 MPa occurring in severely desiccated soils. Under these conditions it is unlikely that roots would be able to penetrate through the soil. Instead they will exploit macropores within the soil, created when the soil cracks as it dries and shrinks, or biopores produced by previous root growth or the tunnelling activity of invertebrates.

It should be apparent from this that some soils, particularly clay soils, are relatively hostile for roots. By comparison, a soil with a well developed structure, a varied particle size, well aerated and with abundant supplies of available water, will provide favourable conditions for roots. These ideal conditions can occur anywhere within the soil volume but are most usually found near the soil surface. It is for this reason that the great majority of roots are found in the surface 600 mm. However, if the opportunistic exploitation of the soil encounters favourable conditions at any location, roots will proliferate.

Although other factors are also important, it is the search for water which is by far the dominant factor in determining the scope of the root system. The internal resistance of roots to water flow is very low. For this reason it can be more efficient for a tree to allocate carbon reserves to an extensive root system exploiting the optimal conditions wherever these are encountered, rather than expending energy on growth in adverse conditions closer to the tree.

Depending on the shape of the crown, there is often an abundance of roots beneath the trunk exploiting the rainfall which tends to run down the trunk of the tree (stemflow), or beneath the drip line on the outer periphery of the crown. By contrast, the crown may intercept rainfall and produce a rainshadow beneath, thereby encouraging roots to spread beyond the crown. Roots will also proliferate near any natural water source, and, if they encounter them, will exploit deep aquifers such as bands of sand within an otherwise hostile clay soil, particularly if there is lateral inflow to replenish the source. In the same way they will exploit artificial sources of water, such as leaking drains, or the moisture condensing on the underside of paving slabs.

It should be appreciated from this that, unlike the leaves which are concentrated on the periphery of the crown, the fine roots will develop in whichever part of the root mass is most suitable. They can develop off roots of any size; they are not restricted to the ends of the roots in a similar manner to the leaves on the ends of branches. If soil and moisture conditions are uniform around the tree there will be uniform exploitation and root growth, but if there is any variation, the root system will be equally variable.

At all stages in root development there is a high mortality. The fine roots may only live for a few weeks, and, just as the leaves are shed each winter, a large proportion of the conducting root system will die back each year. Even the larger structural roots will diminish in number as the tree matures, and as these roots die, all of the smaller roots attached to them will die. It must therefore be recognised that the whole root system is dynamic, with new roots being produced, living for variable times and growing to a variable extent before dying. The root system is not immutable, but will respond to changing conditions.

Relationship between size of crown and root system

The overall size of the root system depends on the size of the aerial parts, and vice versa. The ratio of the dry weights of the two parts, known as the root:shoot ratio, only varies to a limited extent. For most species of tree, it is within the range 1:10 to 1:3 (Table 2), but for some plants with specialised root systems, such as sugar beet, it can be as high as 3:1. The ratio is dependent on the state of maturity of the plant, generally decreasing with age from seedling stage onwards, i.e. the proportion of roots to crown of a tree decreases as the tree matures. It is also dependent on the soil conditions; if water is readily available there will be a lower root:shoot ratio than for a similar tree in a soil of low water availability. The concept of root:shoot ratio is based on dry weight, but in terms of the volume of soil encompassed by the roots, initially root volume is greater than the crown volume but this is reversed as the tree grows.

Table 1. Yearly production of organic matter (oven-dry weight, tonnes/hectare), and associated root:shoot ratios (derived from Bray, 1962).

Tree species	Below ground	Above ground	Total	Root:shoot ratio
Beech (Germany)	1.6	8.2	9.3	1 : 5.1
Birch	2.2	6.7	8.9	1 : 3.0
Scots pine (mean)	1.6	8.9	9.3	1 : 5.6
Norway spruce (Ger.)	2.1	11.9	13.5	1 : 5.7
Rain forest (Ghana)	2.6	21.7	24.3	1 : 8.3
Herbaceous species				
Vetch (U.S.A.)	0.7	4.7	5.4	1 : 6.7
Wheat (mean)	2.0	6.8	8.8	1 : 3.4
Maize (U.S.A.)	4.5	8.7	13.2	1 : 1.9
Potato (mean)	4.0	2.6	6.6	1.5 : 1
Beet (mean)	9.5	3.1	12.6	3.0 : 1

Although the ratio varies through the life of a tree and can be influenced by a change in conditions, for any individual it is a very fundamental value which is under tight control in the allocation of carbon resources. If the ratio is upset for any reason, for instance by damage or by pruning either the root or shoots, the tree will seek to readjust back to the original relationship, either by enhanced growth of the roots or shoots if this can be achieved, or by dieback of tissue which is in surplus. This provides a potential mechanism for controlling root growth, which is considered in a later section.

Although the overall size of the crown and root system are closely linked, as the distribution of the roots within the soil is strongly influenced by even small variations in soil conditions, it is not possible to make reliable deductions about the radial spread or depth of roots. It can be very misleading to think that root spread will equate to canopy spread or to tree height or any other parameter of crown size. In some circumstances, spread will be far less, while in others active roots may be found far beyond their expected location.

Transpiration - the driving force

The reason why roots take water is to replenish that which is lost from the leaves during photosynthesis. Photosynthesis is the essential process of plant growth by which carbon dioxide is converted to sugar using the energy of sunlight. The carbon dioxide must first be absorbed into the cells within the leaf; this occurs in a film of water which covers the enormous surface area of special spongy cells inside the leaf, in a manner similar to absorption of oxygen within our lungs. Inevitably, water will evaporate from the film; this loss of water is called transpiration. Plants have evolved highly efficient mechanism for controlling transpiration, but inevitably, as carbon dioxide diffuses into the leaf, water vapour will be lost. Over 99% of water used by a plant is lost as transpiration, with the amount of water used closely related to the production of sugar, and thus the growth, achieved by the plant.

Plants, whether they are grass or trees, have a leaf area in summer which is up to 6 times greater than the ground surface area they cover. Furthermore, the spongy inside to their leaves creates a massive surface area available for evaporation. The plan can exercise some control of transpiration loss, as the gaseous diffusion occurs through minute pores on the surface of the leaf, known as stomata, which will close if the leaf is under water stress. Even so, transpiration is typically about 6 times greater than the evaporation which can occur from open water, or from a bare soil surface.

As water is lost from the leaf, it sets up a suction within the leaf. The leaf is at the extremity of a continuous system of vessel cells, which extend from the roots, through the trunk, branches and twigs to terminate in the fine veins of the leaf. The loss of water from the leaf sets up a gradient throughout this continuous system, so that water flows from the roots to the leaves.

The suctions which develop are quantified in terms of water potential, and have been extensively studied and quantified (see Kozlowski (1982) for a detailed review). The potential which develops in a leaf can vary widely, with some desert plants and also alpine plants being capable of developing very high suctions (6000 kPa or greater), but in temperate climates maximum suction in the leaf is typically between 1500 and 2000 kPa. Within the branches and main trunk of trees there is a decrease in the suction as a result of the resistance to water flow. This gradient between the leaf and root is about 200 to 500 kPa, so that the maximum suction exerted by the roots on the soil is typically about 1500 kPa. It is the development of this suction which causes water to move from the soil into the plant. 1500 kPa is commonly considered the maximum suction achieved by tree roots in temperate climates at their wilting point.

Where water is freely available, maximum photosynthesis will occur and water flow and loss will be rapid. The rate of this will depend on the weather conditions and the daylength, but peak rates of evaporation from the soil surface, and transpiration from leaves (commonly combined into the term evapotranspiration) in June/July in southern England will increase the soil moisture deficit by more than 5 mm per day. This equates to a loss of 50,000 litres per hectare. If there is a continuous forest canopy, it would be possible to convert this to the water use per individual trees (i.e. at 100 trees/hectare, each would be losing 500 litres/day). An isolated tree will potentially have a far higher water use as it will have a far larger leaf area (extending down the side of the crown to ground level) and may be exploiting soil from a surface area extending far beyond the canopy spread.

In practice, even if water is freely available and conditions are conducive to high transpiration rates (i.e. high temperature, low vapour pressure and sufficient air movement), the absorption of water usually cannot keep pace with transpiration loss and a deficit develops within the plant which may be sufficient to lead to loss of turgidity and temporary closure of the stomata. As a result, there are marked diurnal variations in deficit, with the minimum at or near dawn and a maximum in late afternoon. For instance, the suction in leaves of sugar maple (*Acer saccharum*) and paper bark birch (*Betula papyrifera*) trees increased from about 300 kPa in the morning to as much as 2000 kPa for the maple and 1600 kPa for birch in the early afternoon (Periera and Kozlowski, 1978). This increase in the suction and build-up of a water deficit results in stomatal closure to reduce the amount of water loss. Under conditions of extreme drought, a tree can remain under permanent water stress with the stomata closed and negligible further water loss.

For these reasons, and because it is unusual for water to be freely available, the actual rates of transpiration are far lower than the theoretical maxima. In such circumstances it becomes a meaningless exercise to try to calculate the amounts of water which might be lost from a tree, or to use this as a basis for comparison. Water loss will always be heavily dependent on water availability.

Roots of grass can be extremely efficient at drying the soil down to depths of at least 1m, with some influence extending deeper, but their radial influence is very limited. The suctions generated by tree roots is no greater, but tree roots can extend far deeper and also extend laterally beneath a building, and thus have a far greater zone of influence. However, it must be remembered that tree roots require oxygen for their survival and have very limited ability to penetrate soil with a bulk density greater than 1.8 g/cm^3. These factors severely restrict the amount of root growth of most tree species in clay soils.

Seasonal soil drying

The rates of photosynthesis, and thus transpiration, and also evaporation, are dependent on sunlight, temperature, vapour pressure and wind speed. They reach a maximum in mid summer, and then diminish as the summer wears on.

The extent to which this evapotranspiration will dry the soil depends on the input of water from rainfall. In the UK, during summer months the rainfall is usually less than the rate of evapotranspiration, and so the extent of soil drying will gradually increase, typically reaching a maximum in about September. Figure 2 presents a series of soil moisture profiles which show the gradual development of soil drying through

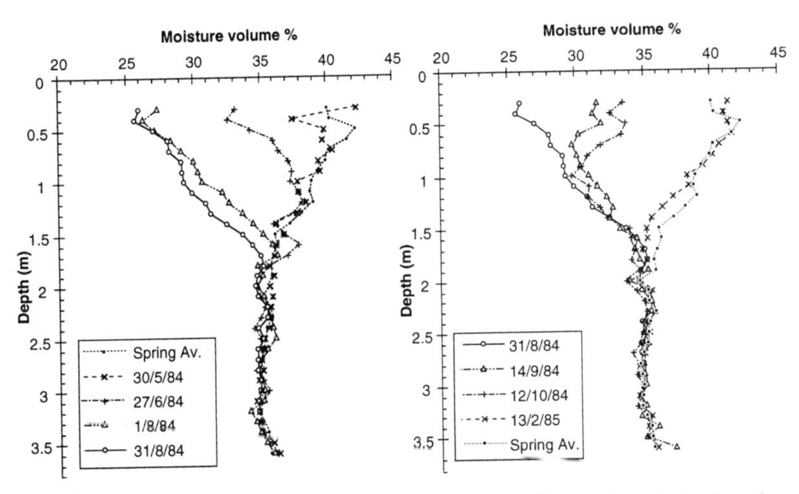

Figure 2. Progressive development of soil drying during summer, 2.8 m from lime tree on Boulder Clay.

Figure 3. Progressive rehydration of same soil during winter.

the summer in close proximity to a lime tree (*Tilia*)[2]. In the autumn, as the rate of transpiration diminishes, the amount of rain is usually greater than evapotranspiration, allowing the soil to start to rehydrate. After leaf-fall there will be negligible water loss from deciduous trees. Evergreens, although maintaining a potential for photosynthesis, will be restricted by low temperature and light levels so that transpiration is minimal. As a result, during the winter the soil will be rehydrating (Fig. 3), until it reaches field capacity when no further water can be absorbed, and run-off occurs.

Using soil moisture profiles as shown in Figs. 2 and 3, one can calculate the soil moisture deficit below any selected depth (Fig. 4 overleaf). Deficits are obviously greatest near the surface, and in this example are negligible below 2 m depth (the maximum depth to which drying occurs). With a clay soil, shrinkage is related to the extent of the soil moisture deficit, and so these curves provide an indication of the pattern of soil shrinkage which will be occurring at the different depths.

The factors which govern the rate of evapotranspiration (sunlight, temperature, vapour pressure) are fairly similar each year, and so the potential rates of evapotranspiration remain fairly similar each year. The greatest variation in climatic conditions is in respect of rainfall, which is therefore the main factor to influence the

[2] The moisture content profiles and measurements of soil moisture deficit presented in Figs. 2 - 8 were obtained by the author as part of a long-term project using neutron probes to monitor changes in soil moisture content in proximity to 60 trees of varying species on a variety of clay soils. Full details will be found in Tree Root Damage to Buildings; Volume 2, Patterns of soil drying in proximity to trees on clay soils (Biddle 1998).

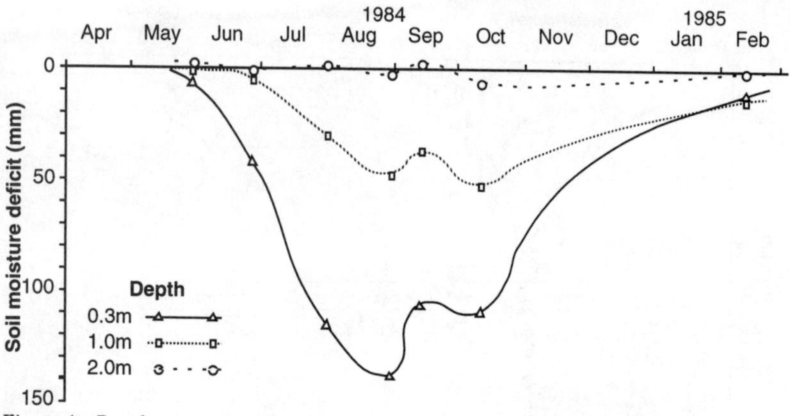

Figure 4. Development of soil moisture deficit at varying depths, derived from data in Figures 2 and 3.

maximum soil drying to develop each year. Figure 5 shows the driest profiles recorded in various years in close proximity to the same lime tree. Even in a wet year, such as 1992, there was significant drying. The long-term seasonal fluctuations of soil moisture deficit in spring and autumn at varying distances from the same lime tree are shown overleaf in Fig. 6.

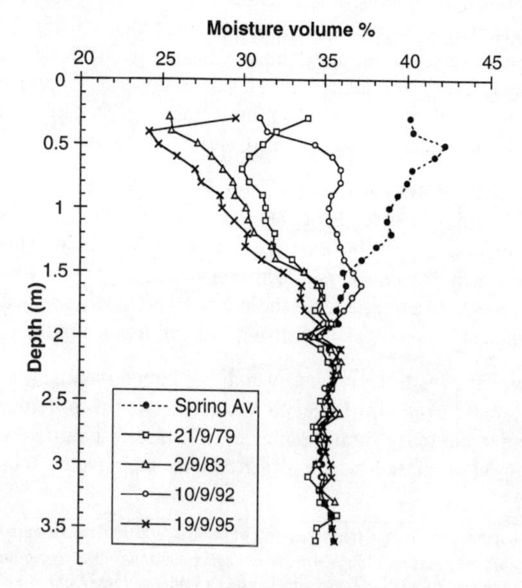

Figure 5. Maximum soil moisture deficits developing in different years, 2.8 m from lime tree on Boulder Clay.

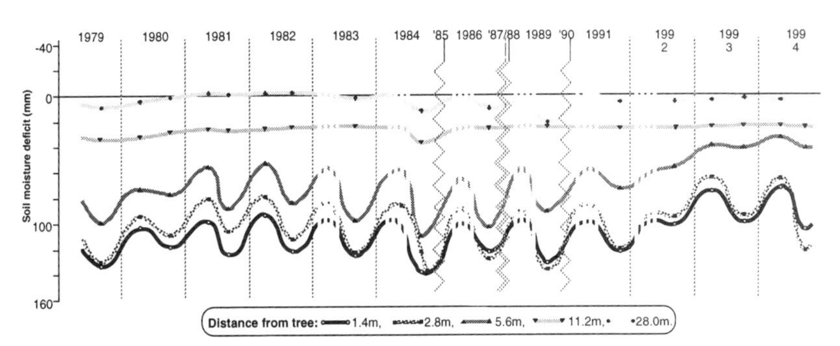

Figure 6. Seasonal fluctuations in soil moisture deficit at varyingdistances from lime tree on Boulder Clay

Even in dry parts of the UK, the average winter rainfall is greater than the soil drying which develops during the summer, even in proximity to a large tree. Unless it is an exceptionally dry winter, there is therefore plenty of rain to correct deficits which develop during the summer, and any influence of a tree or other vegetation is entirely seasonal, with the soil returning to field capacity during each winter. This pattern of seasonal drying applies to the majority of situations where trees are causing structural damage in the UK.

Persistent soil drying

Although there is usually plenty of winter rain in the UK to correct deficits, if the permeability of the soil is too low it can restrict the extent of rehydration, so that the soil is still partially desiccated in the spring when soil drying recommences. Over successive years, if drying is greater than rehydration, the extent of soil drying will increase. Figure 7 shows the moisture content profiles recorded in the spring of various

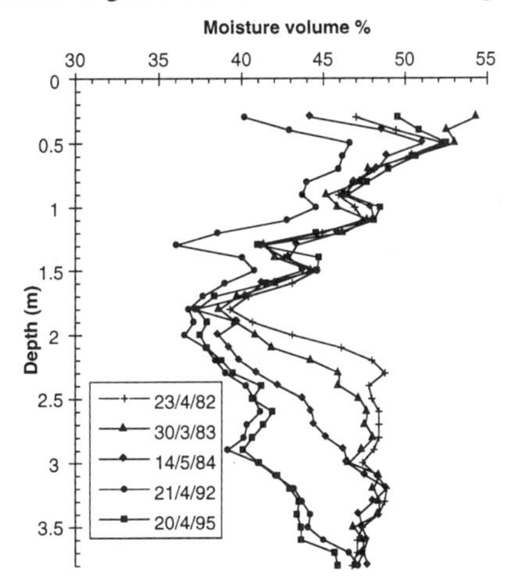

Figure 7. Progressive development of a persistent moisture deficit as result of soil being unable to rehydrate during a single winter. (4.5 m from poplar on London Clay)

years between 1982 to 1995 in proximity to a poplar tree (*Populus*) on London Clay. In each winter the soil rehydrated fully down to about 1.7 m (except in 1992, which was an exceptionally dry winter), but below 1.7 m the soil was becoming progressively drier as a result of drying during the previous summer. Figure 8 shows the progressive increase in soil moisture deficit in proximity to this tree.

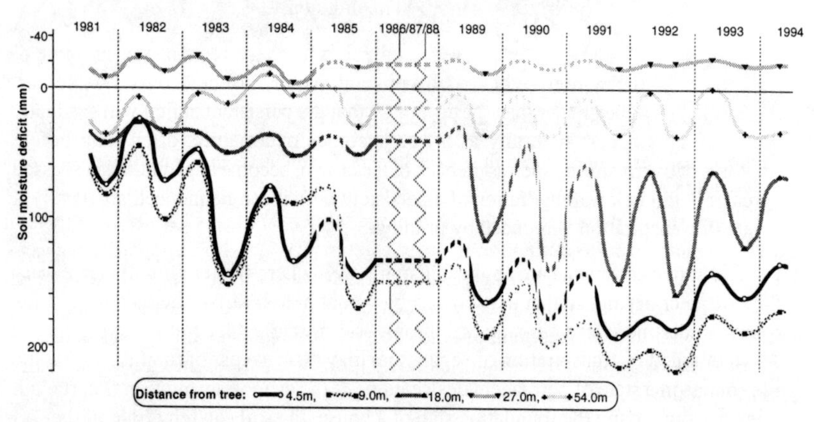

Figure 8. Seasonal fluctuations and progressive development of persistent soil moisture deficit at varying distances from a poplar tree on London Clay.

There are two components to the soil drying which grade imperceptibly into each other:

i) a seasonal effect, controlled by the permeability of the soil which influences the amount of rehydration which can occur in the winter;

ii) an underlying zone of soil which remains persistently desiccated. The extent, both vertically and laterally, of this persistent deficit will depend on the ability of the vegetation to cause soil drying below the depth to which seasonal recovery can occur.

In the UK there is a very common misconception that a persistent deficit is the norm in proximity to most trees on most soils. In practice in the temperate climate with high winter rainfall, they are infrequent, with significant persistent deficits only developing in clay soils of very low permeability and with a few species of tree (such as poplar, elm (*Ulmus*) and oak (*Quercus*)) which are capable of exploiting the very hostile rooting environment of such soils. Clay soils can, of course, have exceedingly low permeability, but under natural conditions the soil structure, with its three-dimensional network of cracks, which are usually lined with slightly coarser and more permeable soil particles, is usually sufficient to allow almost complete recovery each winter.

In those situations where a persistent deficit does develop, research and observations indicate that it establishes over comparatively few years, usually during the period of very rapid growth during early maturity of a tree. As the deficit develops, the soil conditions become ever less suitable, with the situation soon being reached where it is easier for the tree to search elsewhere, or where lack of water restricts the rate of further growth. Once this stage is reached, the persistent deficit will remain so long as the seasonal drying at the surface is maintained.

However, if the seasonal drying diminishes for any reason, the amount of rehydration in the winter (which is a constant factor, affected only by the soil permeability) will be greater than the drying, so that the persistent deficit will gradually diminish. This occurs naturally as a tree becomes overmature, often long before symptoms of overmaturity, such as dieback of the crown, become apparent. As discussed in greater detail below, rehydration of a persistent deficit can be induced artificially if the rate of transpiration is reduced by pruning.

The above comments are applicable in the UK, where winter rainfall is in excess of the summer drying, so that the soil has the potential to return to field capacity each winter. Under more arid climatic conditions, soil drying, either from evaporation on its own or aided by transpiration of vegetation, may be in excess of rainfall, so that the soil remains in a state of permanent desiccation. If the evapotranspiration is curtailed, for instance by laying the foundation slab of a house, the soil can rehydrate; if the soil is clay, this gives rise to the problems of expansive soils. Under these arid conditions, any tree root activity which takes water in excess of the normal rates of evapotranspiration (i.e. that caused by the naturally occurring ground cover) will increase the depth and/or intensity of soil drying, and thus increase the extent of any persistent moisture deficit. Any soil with the ability to undergo long-term expansion has a persistent deficit.

Whatever the cause of soil drying, the extent to which it will swell as it rehydrates will depend on the swelling characteristics of the clay and the depth and intensity of soil drying; the duration of swelling will be governed by the availability of water and the permeability of the clay.

Implications for remedial action

The implications of these patterns of drying in clay soils should be obvious. If soil drying is seasonal, any movement of the clay, and thus movement of the foundations, can be controlled by appropriate action with the tree (or other offending vegetation). Seasonal recovery will be complete by the next spring. Thereafter, provided the tree has been correctly identified and dealt with and there is no other cause of soil drying, there will be no further foundation movement (Fig. 9 overleaf). Superstructure repairs can be undertaken with confidence that similar problems will not recur.

However, if a tree has created a significant persistent moisture deficit, or increased the desiccation of an expansive clay, any action to restrict the water uptake of the tree

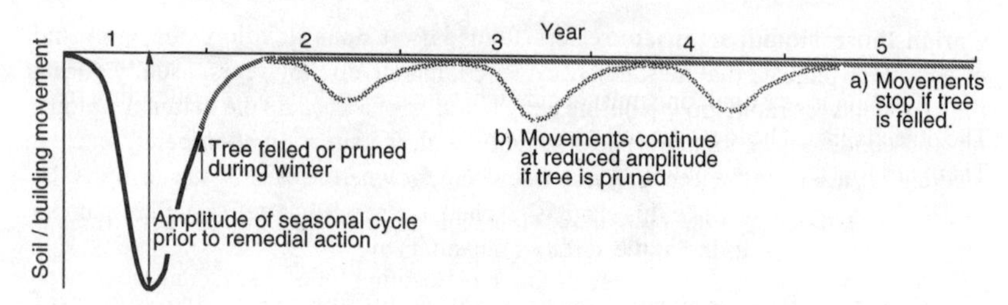

Figure 9. Diagram of remedial options if drying is predominantly seasonal

will allow the soil to rehydrate, causing heave. If the tree is completely felled, the heave will occur progressively (Fig. 10, curve a), but any action with the tree which is sufficient to reduce transpiration will reduce the amount of soil drying each summer, whilst the amount of recovery during the winter will remain the same. As a result, there will be a combination of seasonal movement of reduced amplitude, superimposed on progressive heave (Fig. 10, curve b), so that the amount of movement and associated stress within the building will be increased. If there is a persistent deficit, it is not possible to reduce the amplitude of seasonal movement without also causing heave.

If there is a persistent moisture deficit, unless the swelling movements can be tolerated, underpinning is usually the only practical remedy. Such underpinning should, of course, include full anti-heave precautions to allow for the eventual demise of the tree. In some exceptional circumstances, other techniques, such as accelerated watering-in, may be applicable.

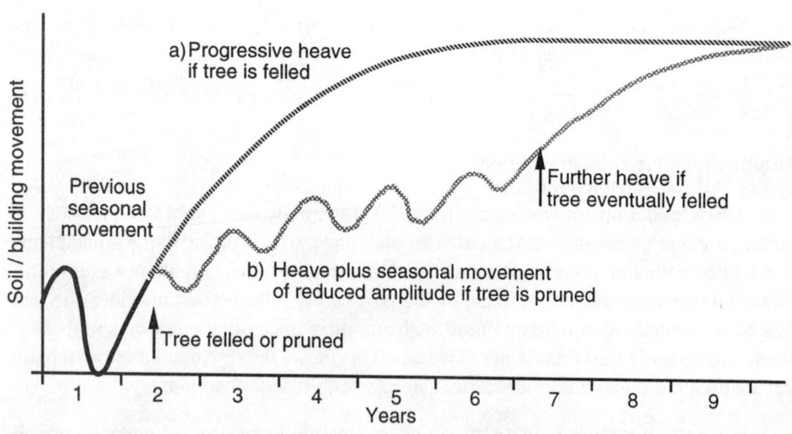

Figure 10. Effect of pruning or felling if there is a significant persistent deficit; stability not achieved.

Options for remedial action to trees.

If drying is seasonal, one must decide what should be done to the offending tree. The objective must be to stop significant root activity by that tree below the foundations. There are four potential options:

i) fell the offending tree to eliminate all future drying (Fig. 9, curve a).

ii) prune the tree to reduce drying and the amplitude of seasonal movement (Fig. 9, curve b).

iii) control the root spread, to prevent drying beneath the foundations.

iv) supplementary watering to prevent the soil from drying.

Tree felling. The most effective method is to fell the tree (or shrub), and in most situations this should be the preferred option. As soon as a tree is felled, all water uptake by the roots will cease. Roots may remain alive for a considerable period, but they will be completely inactive until they eventually die and decay. If the tree is felled in the spring, after seasonal recovery is complete, immediate stability of the foundations will be achieved. If felled in the autumn, the normal period of recovery will be required through the winter, with stability by the spring.

Tree pruning. An alternative is to reduce transpiration by reducing the leaf area, i.e. by pruning the tree. This may not stop all movement, but aims to reduce the amplitude of seasonal movement to an acceptable level. In the UK there is considerable practical experience of the efficacy of this technique when applied to street trees. In many of our cities, potentially large forest species, e.g. plane (Platanus), lime (Tilia) and sycamore (Acer pseudoplatanus) were widely planted in the late 19th/early 20th century, often in very close proximity to buildings with shallow foundations on clay. It was the original policy for these trees to be kept regularly and heavily pruned removing all regrowth every one or two years. As long as this policy was maintained they were kept to a small size, but in the early 1970's there was a change in policy and pruning was stopped, partially as a result of cut backs in local authority expenditure but also in response to demands to allow the trees to develop a more natural appearance. As a result the trees developed a progressively larger crown, with larger leaf area. The larger leaf area enabled more photosynthesis, with this reflected in increased wood production and also inevitably in increased transpiration. The increase in wood production can be demonstrated by taking a core of wood from these trees and measuring the width of the annual growth rings. Ideally one needs to measure the amount of wood laid down on all the trunk, branches, twigs and roots, but it is usually sufficient just to determine the radial growth and use this to calculate the increase in area of the trunk each year (Fig. 11 overleaf). This example shows the massive increase in basal area increment (i.e. wood production) since regular pruning was stopped. There will have been a corresponding increase in transpiration, which will have required a considerable increase in root activity. This change in tree management policy resulted in an enormous increase in cases of damage by trees in these situations.

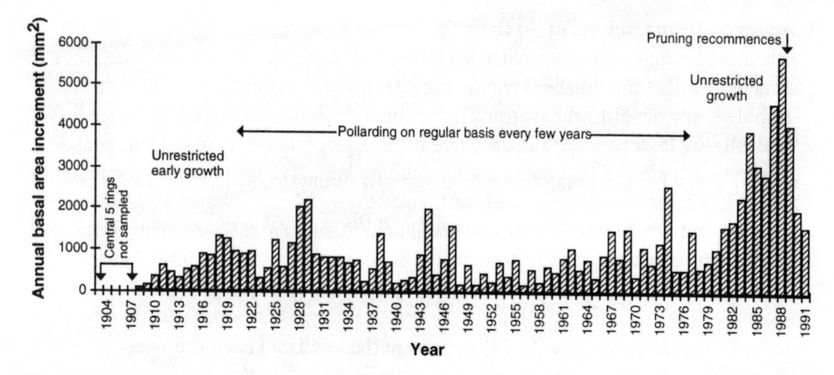

Figure 11. Annual increment growth of a plane tree subject to varying pruning regimes durings its life.

Research and experience show that, if the previous policy of heavy reduction is reinstated, the tree will revert to its previous water requirements, and the root system will eventually die back to a size commensurate with the smaller crown size. Likewise, even if a tree has not previously been pruned, treatment can be started so as to reduce leaf area and transpiration.

However, there are many potential problems. In particular, if damage has already occurred, it is usually necessary to reduce water uptake quite drastically if the remedy is to be effective. The extent of essential work will very often spoil the appearance of the tree, and so it is understandable that the necessary work may be resisted by the tree owner. However, unless sufficient leaf area is removed, the treatment will be ineffective. After they are pruned, most species will respond with vigorous new growth to re-establish their root:shoot ratio. It is often mistakenly suggested that this indicates enhanced vigour and thus greater water uptake, but this growth is achieved using food reserves rather than from photosynthesis, and so does not involve extra transpiration. However, regrowth can soon restore the original leaf area, and it is therefore necessary to keep pruning the tree on a regular basis if the leaf area is to be kept under control. One of the reasons why plane and lime were chosen for planting in city pavements was because they are reasonably amenable to regular pruning, and because this was an effective management regime in those circumstances. Many other species respond far less satisfactorily, either failing to produce any regrowth (e.g. mature beech (*Fagus*), or many species of conifer) or with excessive or untidy regrowth (e.g. poplar or cherry (*Prunus*)). If it is necessary to prune a tree severely, it will usually create large wounds which may allow entry of decay and create problems in the future. For all of these reasons, pruning is only appropriate in some circumstances, and certainly does not provide a universal panacea. Felling should usually be the preferred option.

If trees are to be pruned, there are various techniques, either involving reducing the overall size (crown reduction) or else thinning the density of branches whilst maintaining a similar outline (crown thinning). Research is currently underway at Horticulture Research International in the UK to determine the efficacy of these different treatment. Initial (unpublished) results suggest that crown thinning by 30% produces negligible if any benefit, whilst crown reduction by a similar amount will reduce water uptake, but to a lesser extent than the percentage reduction in leaf area. Thus, peak soil moisture deficits in the first autumn after pruning were 112 mm beneath the reduced trees and 135 mm beneath the control and thinned trees. This difference was statistically significant, but would be of limited practical benefit, and emphasises that very severe crown reduction may be required to achieve an effective treatment. The difference in behaviour between the thinned and reduced trees is attributed to behaviour of the boundary layer around the leaf. On the crown thinned trees, the open canopy produces a minimal boundary layer, so that the fewer leaves tend to lose more water than the controls. By contrast, the reduced trees have a small but compact crown size. Individual leaf size is increased, but these leaves are shaded and screened from the wind, and so have a thick boundary layer which reduces the rate of transpiration.

Root restriction. Controlling root growth might appear a more attractive proposition than controlling the leaf area, but there are usually practical problems.

Simple root severance can be effective, provided all roots extending beneath the foundations can be treated without detriment to tree stability and safety. However, the variable root distribution makes it difficult to know whether all roots have been severed, and the roots will soon regrow. Repeat treatments on a regular basis would be essential, but there is a high risk that the need for these would be out of sight and thus out of mind.

Physical root barriers are sometimes advocated, but again usually create insuperable problems. The barrier must be sufficiently far from the tree not to affect stability, must be impenetrable, and must extend to sufficient depth and lateral extent to ensure that there is no risk of roots passing through, under or around the barrier. It must also extend above ground level to prevent roots growing over the top. As the soil on the side nearer the tree might be subject to seasonal movements of 100 mm or more while the soil on the far side should be stable, there can be practical problems in integrating a barrier into the landscape. These soil movements also make it difficult to seal around any underground services.

If a barrier is installed, various materials can be used, but either thick polythene sheet (at least 1 mm) or a thick geotextile membrane are the most common. Geotextiles can have the advantage of allowing moisture movement, whilst preventing root penetration. Various proprietary membranes, incorporating slow release herbicides, are also available.

Supplementary watering. In theory it might be possible to prevent the soil from drying by ensuring adequate water supply at all times. However, the quantities of water required by the tree usually make this impractical, particularly as water may not be available when it is most needed during periods of drought.

Conceivably there could be an advantage in providing suitable soil conditions and a regular supply of water in an area well away from the foundations, in order to encourage the tree to root away from the building. Provided adequate water is provided, soil drying in other areas, such as beneath the foundations, would be reduced. However, if the water supplies dry up, the tree would start to take water from other sources where it is available, immediately putting the building back at risk.

A leading drain can be considered a source of supplementary water. If it is repaired, it can lead to changes in the pattern of soil drying and a risk of structural damage.

The investigation of tree root damage

The objectives of subsidence investigations are twofold:

i) to determine the cause;

ii) to determine the appropriate remedy.

In the UK, far too often investigations are aimed solely at determining the cause, particularly if a neighbour's tree is suspected, as the neighbour's insurers may then be liable for the cost of repairs. Traditionally these investigations concentrate on digging trial pits to identify roots below the foundations, and boreholes to try to demonstrate that the soil is desiccated. This information is of limited value for determining the remedy.

Root identification is of little practical relevance. Even if fine roots are present, it does not prove that they are taking sufficient water from the soil to cause shrinkage, whilst roots which are large enough for identification are merely conducting the water from some available and possibly distant source, not necessarily from below the foundations. If the roots of more than one species may be present, one must take sufficient samples to identify all the trees, but their relative abundance is not necessarily a reliable indication of their contribution to any soil drying.

In my opinion far too much emphasis is usually placed on soil investigations, as interpretation of the information is fraught with problems. If a soil is severely desiccated, there is usually no ambiguity; for this reason boreholes drilled in the autumn can usually demonstrate desiccation if trees are involved. However, this is of limited practical value, as decisions on the type of remedial action should depend on whether any such desiccation is seasonal or persistent. This information can only be obtained from boreholes drilled in the spring, as it is only at that time of year that one can assess whether the soil has recovered fully (i.e. that drying is seasonal), of whether the drying is persistent. Most of the methods of determining desiccation are inadequate for identifying marginal levels of desiccation which can still be sufficient to allow serious heave. Even if the depth and extent of desiccation can be established, methods for quantifying the potential swelling of the clay are crude and inaccurate.

If boreholes are being drilled, assessment of the permeability of the soil is the most relevant information to obtain, as it is this which determines the duration of recovery. Practical methods are not available for measuring permeability, but with experience it can be assessed on the basis of macro and micro soil features. However, I have yet to see any engineering report which provides any such assessment.

I am not denying that soil investigations have a role, but, before embarking on the work, the objectives and potential value of the information should be identified. Trial pits to determine foundation depth may be needed, if only to check on the adequacy of foundations and their vulnerability to shallow rooted vegetation, and to check whether clay is present. In some situations more detailed borehole investigations may be appropriate, and will certainly be required if there is a persistent deficit or if underpinning is required because of some other cause. However, a stereotyped approach to all investigations, with an automatic request for trial pits and boreholes regardless of the time of year of other features of the case, demonstrates a lack of understanding and awareness of the problems of tree root damage.

Level monitoring - the essential part of investigations

If trees are the cause of damage, a highly diagnostic feature is the pattern of seasonal movement, with the foundations subsiding each summer and recovering in the winter. By contrast, other causes of subsidence produce a progressive downward movement, whereas heave will produce a progressive upward movement. If damage is caused by movements of the superstructure, there may be active crack movement but the foundations will be stationary. It should therefore be obvious that one of the most effective ways to distinguish between the causes is to study the pattern of foundation movement. This can be done easily, quickly and accurately by precision level monitoring.

If this demonstrates seasonal movement and thus the involvement of vegetation, it is then necessary to identify the offending tree or shrub. Level monitoring will provide this essential information, as it will show exactly which parts of the building are moving, and by how much. Movements will usually be greatest closest to the offending tree, unless there are abnormalities in the soil or variations in foundation depth. If the movements are widespread, they imply a large tree with an extensive influence, whereas localised movement suggests a smaller and closer tree or shrub.

Even if a deep datum is not available, it is usually not difficult to establish a reliable stable datum, as in most situations only parts of a building will be moving, with the direction being downwards in summer or upwards in winter if trees are involved. The marker which best demonstrates this pattern on the rest of the building can usually be assumed to be stable.

Once the offending tree is identified, and before remedial action is taken, one needs to determine whether movements are seasonal, or whether there is a persistent deficit. Again, level monitoring is one of the best methods for obtaining this information. Tree root damage typically occurs in the late summer/autumn. If monitoring starts as soon as a claim is notified, it can record the pattern of recovery movement through the winter. If the rate of movement slows and becomes negligible by spring, it indicates that recovery is nearly complete, i.e. that movements are seasonal and that vegetation control will be effective. However, if the rate of movement remains steady, it indicates a potential for it to continue, because of rehydration of a persistent deficit. Unless this continuing movement can be tolerated, there will be a need to underpin.

There is a common misconception that level monitoring is too lengthy an operation and will delay remedial action. Apart from a brief period in spring and autumn when the direction of movement is reversing, a set of readings taken 6 weeks after initial set up will distinguish those parts of the building which are moving, and thus identify the probable culprit. As soon as the culprit is identified, arrangements can start for necessary pruning or felling, with subsequent readings used to confirm the diagnosis before action is taken. If monitoring can start in the autumn, two or three subsequent sets of readings through the winter will determine changes in the rate of movement and thus whether there is a persistent deficit. This should also provide sufficient evidence to substantiate any demands for action on trees which are owned by a third party. If one is to stop a further cycle of seasonal movement and subsidence, it is then essential that decisions are finalised and action taken with the culprit in the spring. Even if monitoring starts at other times of year, for instance in the spring, it can detect the initial movements and allow action before further damage occurs in autumn.

Level monitoring should not cease when the tree is felled or pruned, but continue so long as needed to confirm that the remedial action has been effective. This is particularly necessary if the tree has only been pruned, to check that the movements have been reduced sufficiently (measurement of the amount of recovery prior to any pruning provides a useful basis for comparison).

Despite the enormous advantages of level monitoring, in the UK crack monitoring still remains far more common. However, it does not begin to provide comparable information. At best it will show that movements are seasonal but must usually span a far longer period before reliable conclusions can be reached. Even then crack monitoring will not distinguish whether these are caused by seasonal movements of the foundations, or, for instance, moisture content changes in superstructure timbers. More importantly, crack monitoring will not show which parts of the building are moving, and is of no use for heave prediction. Its main, or only, use should be for confirming that action to stabilise the foundations has been effective in stopping crack movement.

It is sometimes suggested that level monitoring requires two operatives and is therefore more expensive than crack monitoring. If a lot of monitoring is being done, a team of two can work faster, but simple methods exist to allow easy single-handed operation. Level monitoring must however be accurate and reliable, with readings taken to the nearest 0.1 mm. The use of a heavy staff, resting on a bendy nail (as is seen far too often) is woefully inadequate. Working single-handed, I can take about 12 readings around a typical detached house in less than 40 minutes, usually with a closing error of less than 0.5 mm. This is certainly slower than monitoring crack widths, but as the information provided is so much more useful, the extra costs become insignificant

A level distortion survey of a brickcourse around a building or of the internal floor slab (as distinct from level monitoring) can also provide useful information on the movements which have occurred sometime in the life of the building (but not necessarily recently). If trees are involved, distortion will normally point towards the culprit.

It is helpful to have the results of initial level monitoring , or of a level distortion survey, before deciding on the location of any boreholes or trial pits. This helps to ensure that soil information is obtained from the most relevant location.

Assessing the influence of trees

Identifying which tree(s) or shrub(s) are the cause of damage is essential before effective remedial action can be taken. As previously noted, the distribution of movement within a building which is recorded by the level monitoring will usually provide a good indication on the most likely culprit. In many cases this evidence alone will be sufficient, but additional assessment of the potential influence of particular trees or shrubs may be necessary.

It is widely recognised that some species of tree are more likely to cause damage than others. Different lists have been produced by various authors in the past; the differences between these may reflect evolving knowledge, or the use of different criteria as the basis for assessment (e.g. radial root spread, depth of roots, propensity to cause damage, or 'water demand'). 'Water demand' is a term which is commonly used, but in biological terms it is a misnomer, as the demand for water, or suction, exerted by most plants will be similar. However, if 'water demand' is defined as "the ability of vegetation to cause drying of a clay subsoil", it provides a useful concept. Table 2 summarises my current classification, based on the 'water demand' of tree genera which are common in the UK.

Table 2. A tentative classification of the 'water demand' of different tree genera in the UK (Biddle, 1998).

Highest water demand (deepest and furthest extent)	- - - - - - - - - - →				Lowest water demand (shallowest and least extent)
Broad-leafed genera					
Eucalyptus	*Crataegus*	*Aesculus*	*Acer*	*Ailanthus*	*Catalpa*
Populus	*Salix*	*Fraxinus*	*Castanea*	*Alnus*	*Corylus*
Quercus	*Sorbus* (simple-leafed)	*Platanus*	*Fagus*	*Betula*	*Ficus*
	Ulmus	*Tilia*	*Malus*	*Carpinus*	*Liquidambar*
			Prunus	*Gleditsia*	*Liriodendron*
			Pyrus	*Ilex*	*Magnolia*
			Robinia	*Juglans*	*Morus*
			Sorbus (compound-leafed)	*Laburnum*	*Sambucus*
Coniferous genera					
Cupressus	*Chamaecyparis*	*Sequoiadendron*	*Cedrus*	*Juniperus*	*Abies*
	x Cupressocyparis		*Thuja*	*Taxus*	*Araucaria*
				Tsuga	*Ginkgo*
					Larix
					Picea
					Pinus

However, it must be emphasised that the influence of trees is very variable and unpredictable. There is likely to be considerable genetic variation between individuals within each genus, and the condition of the tree is likely to be as, or more, important than its genus. A vigorous healthy tree in early maturity of a species of inherently low 'water demand' may be having a greater influence than a larger mature specimen, even if the latter normally has a higher 'water demand'. Soil conditions, particularly under urban conditions, will also be very variable, producing very variable rooting patterns. For these reasons, guidelines based on 'zones of influence', or other criteria, should only be used as an aid to interpretation of level monitoring results, and never used on their own to determine which tree might be the cause of damage.

It is sometimes necessary to determine the age of any tree, and any previous history of pruning, to correlate with any previous evidence of foundation movement. This information can be determined from cores of wood taken with an increment borer, as in Fig. 11.

Prediction and prevention of damage.

With new building, properly engineered foundations should be capable of preventing any risk of future damage. If there are existing trees, one can design for their anticipated effects on the soil. Alternatively foundation design can make provision for future reasonable tree planting. The type of foundation which is adopted is likely to be influenced by other design criteria. For instance, in the UK most properties are built with comparatively shallow (1 m) narrow strip foundations and a ground bearing floor slab. Builders are loathe to depart from established practice, and so if trees are present they merely adapt these foundations by increasing their depth, sometimes involving strip footings up to 3.5 m deep. This is an ineffective way of dealing with the potential problem, and it is slowly becoming accepted that a pile and raft foundation is both cheaper and more effective. If properly designed foundations of this sort became the norm, the problems of damage would eventually disappear.

However, we must also deal with the existing housing stock, much of which has shallow foundations on highly shrinkable clay soils. Even with these very vulnerable buildings, it is often noted that trees in apparently high risk situations have not caused damage, whilst conversely a tree may cause damage in situations where the risk appears minimal. This merely illustrates that the development of damage is very variable and unpredictable.

If it were possible to identify risk situations with reasonable accuracy, there would be every justification in trying to eliminate any tree which is deemed an unacceptable threat. However, in the UK at least, it is my opinion that accurate prediction is not possible, and that it is environmentally unacceptable and economically wrong to try to do so. If all trees which might pose a risk are removed or kept pruned, it would involve massive damage to the urban tree population, and the cost of tree pruning would far exceed the cost of repairing damage.

Instead, it is better to accept that damage will sometimes occur, and to concentrate efforts on quick and accurate diagnosis which will enable properly targeted treatment of the offending trees. Provided damage is dealt with quickly and effectively, repair should usually only entail cosmetic decoration, perhaps with minor stitching of brickwork. If subsidence is dealt with in this way, it will again be accepted as a normal occurrence, rather than a catastrophe which requires the intervention of insurers and expensive underpinning work.

The way ahead

It is important to ensure that the methods of investigation provide a fast and effective diagnosis of the cause of damage. If trees are involved, the investigations must identify the offending tree and determine whether remedial action with the tree(s) will be effective. Accurate level monitoring is by far the best method to achieve this, and should become an essential part of all investigations where the involvement of trees is suspected. Other investigations will only be necessary if some other specific information is needed.

If offending trees can be accurately targeted and dealt with rapidly (before the next growing season), the extent of damage and cost of remedial work will be minimised. This will eliminate the stigma associated with subsidence, and help to overcome the public's concern about trees, which is quite unjustified in most situations.

Bibliography

Tree Root Damage to Buildings. P.G. Biddle (1998) (Volume 1 - 'Causes, Diagnosis and Remedy'; Volume 2 - Patterns of soil drying in proximity to trees on clay soils'). Willowmead Publishing Ltd, Willowmead, Wantage, OX12 9JA, UK. Tel: 00 44 1235 762478.

References

Bray, J.R. (1962). Root production and estimation of net productivity. Canadian J. Bot. **41**, 65 - 71.

ISE (2000). Institution of Structural Engineers report on 'Subsidence of low-rise buildings.' 2nd Edition. 175pp.

Kozlowski, T.T. (1982). Water supply and tree growth. Forestry Abstracts **43**, (2), 57 - 95.

Periera, J.S. and T.T. Kozlowski (1978). Diurnal and seasonal changes in water balance of *Acer saccharum* and *Betula papyrifera*. Physiologia Plantarum **13**, 289 - 299.

Prediction of Volume Change in an Expansive Soil as a Result of Vegetation and Environmental Changes

D. G. Fredlund[1]
and
V. Q. Hung[2]

Abstract

This paper presents a numerical model to predict the volume change in an expansive soil as a result of vegetation and environmental changes. The model formulation is based on the general theory of unsaturated soil behavior. Three typical volume change situations associated with water uptake by tree roots and water infiltration into soils due to watering at surface are presented.

Introduction

Small buildings constructed in expansive soils are often subjected to severe distress subsequent to construction, as a result of changes in the surrounding environment. Changes in the environment may occur as a result of water uptake by vegetation, removal of vegetation and the excessive watering of a lawn. The magnitude of damage to an engineering structure caused by vegetation depends on a series of factors, which includes type of vegetation, soil and groundwater conditions, climate, foundation types and the distance from vegetation.

Many methods of volume change prediction have been proposed. The methods of volume change prediction are based either on soil suction measurements (or estimations) or on one-dimensional oedometer tests. Problems in engineering practice related to volume change and heave, are generally two-dimensional or three-dimensional in nature and are generally due to environmental changes or the effect of vegetation.

This paper presents a finite element, numerical model for the simulation of typical two-dimensional volume change problems that are often encountered in

----

[1]Professor Emeritus of Civil Engineering, University of Saskatchewan, 57 Campus Drive, Saskatoon, SK, S7N 5A9 Canada; phone 306-966-5283; D.Fredlund@engr.usask.ca
[2]Ph.D. Candidate, Department of Civil Engineering, University of Saskatchewan, 57 Campus Drive, Saskatoon, SK, S7N 5A9 Canada

engineering practice. The model formulation is based on the general theory of unsaturated soil behavior. The soil suction conditions in a soil profile are computed through the use of a steady state or a transient unsaturated soil water flow analysis. The volume change constitutive relations for the unsaturated soil are formulated using two stress state variables; namely, $(\sigma-u_a)$ and (u_a-u_w), where σ is total normal stress, u_a is pore-air pressure and u_w is pore-water pressure. Darcy's law is applied to the flow of water, where the coefficient of permeability is written as a function of soil suction. The volume change model is based on two sets of non-linear constitutive relations, (i.e., one for changes in void ratio and the other for changes in water content). While the analysis and example problems are two-dimensional, the theory can readily be extended for 3-dimensional analyses.

The mechanism is explained whereby distress is caused to an engineered structure. The distress is generally related to changes in the ground surface flux boundary condition or due to the water demands of vegetation such as the root system of large trees. The root system of large trees is simulated as a flux boundary condition applied to the finite element mesh. In this way, the volume changes in the surrounding soil mass can be computed.

The engineering problems associated with unsaturated soils have proven to be complex since the soil properties are non-linear functions of stress state variables. The coefficient of permeability and the water storage term for the soil are assumed to be functions of soil suction in the analysis of water seepage. The unsaturated soil volume change functions are determined from typical one-dimensional oedometer results. The elastic modulus functions, E, with respect to changes in net normal stress, and H, with respect to changes in matric suction, are calculated from the swelling indices, C_t, with respect to changes in net normal stress, and C_m, with respect to changes in matric suction, respectively. The Poisson's ratio is assumed to be a constant.

Solutions for the water flow part of the analysis, as well as the stress-deformation part of the analysis, are obtained through the use of a general, partial differential equation solver, PDEase2D (1996). Three typical two-dimensional volume change problems associated with expansive soils are used to demonstrate the applicability of the proposed model. The first situation is associated with deformations in a soil profile due to water being extracted from a root zone (e.g., the roots of trees) within a soil mass. The root zone is simulated as a flux applied to the soil mass. The second situation is associated with the settlement of a house induced by a line of trees growing close to the house. The third situation is associated with the heave of a concrete slab on grade due to loading and water infiltration into the soil.

Theory of Volume Change of Unsaturated Soils

Continuity Requirements

The continuity requirement for an unsaturated soil, assuming the soil particles to be incompressible, can be stated as follows (Fredlund and Rahardjo, 1993):

$$\frac{\Delta V_v}{V_0} = \frac{\Delta V_w}{V_0} + \frac{\Delta V_a}{V_0} \tag{1}$$

where: V_0 = initial overall volume of soil element, V_v = volume of soil voids, V_w = volume of water, and V_a = volume of air.

By using a Cartesian coordinate system and referencing deformation to an elemental volume, the total volumetric deformation of an unsaturated soil element, $d\varepsilon_v$, can be written as the sum of the normal strains:

$$d\varepsilon_v = \frac{dV_v}{V_0} = d\varepsilon_x + d\varepsilon_y + d\varepsilon_z \tag{2}$$

where: $d\varepsilon_x$, $d\varepsilon_y$, $d\varepsilon_z$ = normal strain components in x, y, and z-direction, respectively.

Constitutive Relations

Two stress state variables are needed to describe volume change behavior of an unsaturated soil (Fredlund and Morgenstern, 1977). These stress state variables are net normal stress, $(\sigma\text{-}u_a)$, and matric suction, $(u_a\text{-}u_w)$, where σ is total normal stress, u_a is pore-air pressure and u_w is pore-water pressure. With the use of these two stress state variables, volume changes in the soil due to externally applied loads and the environmental changes (i.e., change in groundwater table, water uptake by a tree or infiltration) can be considered separately.

Assuming the soil behaves in an incrementally isotropic, linear elastic material, the soil structure constitutive relations can be written as follows (Fredlund and Rahardjo, 1993):

$$d\varepsilon_{ij} = \frac{1+\mu}{E} d(\sigma_{ij} - u_a) - \frac{\mu}{E} d(\sigma_{kk} - 3u_a)\delta_{ij} + \frac{d(u_a - u_w)}{H}\delta_{ij} \tag{3}$$

where: ε_{ij} = components of the strain tensor for the soil structure, σ_{ij} = components of the total stress tensor for the soil structure, $\sigma_{kk} = (\sigma_{11} + \sigma_{22} + \sigma_{33})$, δ_{ij} = the Kronecker delta, μ = Poisson's ratio, E = modulus of elasticity for the soil structure with respect to a change in net normal stress, and H = modulus of elasticity for the soil structure with respect to a change in matric suction.

Assuming water is incompressible, the water phase constitutive relations can be formulated in a semi-empirical approach as follows (Fredlund and Rahardjo, 1993):

$$\frac{dV_w}{V_0} = \frac{1}{E_w} d(\sigma_{ii} - 3u_a) + \frac{1}{H_w} d(u_a - u_w) \tag{4}$$

where: E_w = water volumetric modulus associated with a change in net normal stress and H_w = water volumetric modulus associated with a change in matric suction.

Equations 3 and 4 present the constitutive relationships in elasticity forms. These elasticity forms can be used to solve non-linear elastic volume change problems associated with unsaturated soils in two- or three-dimensions. Fredlund and Rahardjo (1993) also presented the constitutive relationships for soil structure and water phase in compressibility forms as follows:

$$\frac{dV_v}{V_0} = m_1^s d(\sigma_{mean} - u_a) + m_2^s d(u_a - u_w) \tag{5}$$

$$\frac{dV_w}{V_0} = m_1^w d(\sigma_{mean} - u_a) + m_2^w d(u_a - u_w) \tag{6}$$

where:

m_1^s = coefficient of volume change with respect to net normal stress [i.e., $3(1-2\mu)/E$];
m_2^s = coefficient of volume change with respect to matric suction [i.e., $3/H$];

m_1^w = coefficient of water volume change with respect to net normal stress [i.e., $3/E_w$];
m_2^w = coefficient of water volume change with respect to matric suction [i.e., $1/H_w$]; and
σ_{mean} = mean net normal stress [i.e., $(\sigma_x+\sigma_y+\sigma_z)/3$].

The constitutive relationships for the soil structure and water phase of an unsaturated soil can be presented graphically (Fig. 1) by plotting void ratio and volumetric water content against the independent stress state variables, $(\sigma-u_a)$ and (u_a-u_w). Coefficients of volume change corresponding to the unloading surface can be subscripted with an "*s*" to represent the word "*swelling*" (i.e., m_{1s}^s and m_{2s}^s).

The volumetric water content versus matric suction relation is called a soil-water characteristic curve (SWCC), and the coefficient of water volume change with respect to matric suction (i.e., m_2^w) is also called a water storage coefficient.

The constitutive surface for the soil structure can also be obtained when void ratio is plotted with respect to the logarithms of the stress state variables (Fig. 2). The

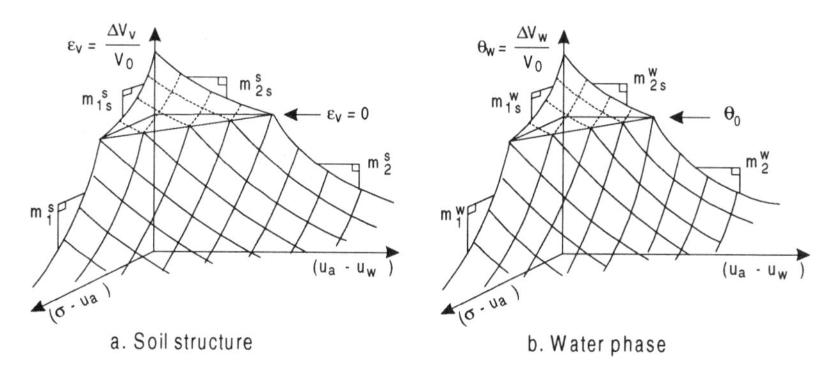

Figure 1. Three-dimensional constitutive surfaces for soil structure of an unsaturated soil (Fredlund and Rahardjo, 1993)

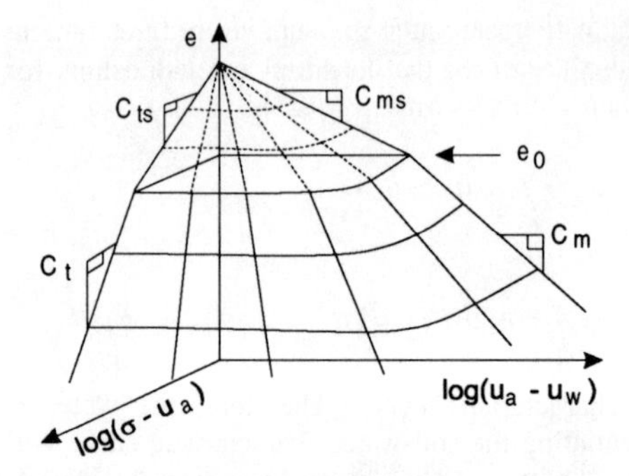

Figure 2. Semi-logarithmic plot of void ratio versus net normal stress and matric suction (Fredlund and Rahardjo, 1993)

logarithmic plots are essentially linear over a relatively large stress range on the extreme planes (i.e., the $log(\sigma\text{-}u_a) = 0$ plane and $log(u_a\text{-}u_w) = 0$ plane) (Fredlund and Rahardjo, 1993). Slopes of the void ratio versus logarithm of net normal stress or matric suction lines are called volumetric change indices, C_t or C_m (Fig. 2). These indices are used in many methods of one-dimensional heave prediction. The elastic modulus function, E is calculated from the volume change index C_t, with respect to changes in net normal stress. The elastic modulus function, H is calculated from the volume change index C_m, with respect to changes in matric suction.

Saturated/Unsaturated Water Flow

In stress-deformation analyses, displacements are calculated from changes in stress states (i.e., changes in net normal stress and matric suction) and the elastic moduli. Matric suction profile is necessary to describe the initial and final stress conditions in soils. If the pore-air pressure is assumed to be atmospheric, the distribution of pore-water pressure is equivalent to the matric suction distribution. The prediction of pore-water pressure must take into consideration in response to changes in the surface flux boundary conditions (i.e., infiltration, evaporation, and evapotranspiration) and the fluctuation of the ground-water tables. The pore-water pressure distribution within the soil can be estimated by performing a saturated/unsaturated seepage analysis.

The governing partial differential equation for water flow through a heterogeneous, anisotropic, saturated/unsaturated soil can be derived by satisfying conservation of mass for a representative elemental volume, assuming that flow follows Darcy's law with a non-linear coefficient of permeability. If it is assumed that the total stress remains constant during a transient process and that pore-air pressure is atmospheric, the differential equation can be written as follows for the two-dimensional transient case:

$$\frac{\partial}{\partial x}\left(k_x \frac{\partial h}{\partial x}\right)+\frac{\partial}{\partial y}\left(k_y \frac{\partial h}{\partial y}\right)=m_2^w \gamma_w \frac{\partial h}{\partial t} \qquad (7)$$

where: h = total head (i.e., pore-water pressure head plus elevation head); k_x and k_y = coefficient of permeability of the soil in the x- and y-direction, respectively; γ_w = the unit weight of water (i.e., 9.81 kN/m^3), m_2^w = the slope of the soil water characteristic curve. Both the coefficient of permeability and coefficient of water storage are dependent on stress states in soils (i.e., net normal stress and matric suction). However, these coefficients of an unsaturated soil are predominantly a function of the matric suction.

The water storage indicates the amount of water taken or released by the soil as a result of a change in the pore-water pressure and can be represented by the slope of the soil-water characteristic curve. Therefore, the water storage function is obtained by differentiating the soil-water characteristic curve with respect to matric suction. Numerous equations have been proposed to simulate the soil-water characteristic curve (Gardner, 1958; van Genuchten, 1980; Fredlund and Xing, 1994). The soil-water characteristic curve described in the present study is limited to the Fredlund and Xing (1994) equation. The Fredlund and Xing (1994) equation is shown below:

$$\theta = \theta_s C(\psi)\left[\frac{1}{ln\left(e + (\psi/a)^n\right)}\right]^m \tag{8}$$

where: $C(\psi)$ = a correction factor for high soil suctions, defined as:

$$C(\psi) = \left(1 - \frac{ln(1 + \psi/\psi_r)}{ln(1 + 1000000/\psi_r)}\right)$$

where: ψ = soil suction (kPa), e = natural log base, 2.71828..., θ_s = volumetric water content at saturation, ψ_r = total suction corresponding to the residual water content θ_r, a = a soil parameter which is related to the air entry value of the soil (kPa), n = a soil parameter which controls the slope at the inflection point in the soil-water characteristic curve, and m = a soil parameter which is related to the residual water content of the soil.

There are many permeability functions that have been proposed to represent the permeability function of an unsaturated soil (e.g., Gardner, 1958; Phillip, 1986; Fredlund and Xing, 1994; Leong and Rahardjo, 1997). These equations involve finding best-fit parameters, which produces a curve that fits the measured data. The equation proposed by Leong and Rahardjo (1997) is used to describe the permeability function for transient water flow analysis in this paper. Leong and Rahardjo (1997) illustrated that the coefficient of permeability is a power function of volumetric water content. Using the Fredlund and Xing (1994) equation with $C(\psi)$ = 1, the permeability function was shown to take the following form:

$$k = k_s\left[\frac{1}{ln\left(e + (\psi/a)^n\right)}\right]^{mp} \tag{9}$$

The parameter p can be determined by using a curve fitting of the coefficient of permeability data. The slope of soil-water characteristic curve (i.e., coefficient of water storage) is obtained by differentiating the Fredlund and Xing (1994) equation (Fredlund, 1995). The water storage function is shown below for correction factor $C(\psi) = 1$:

$$m_2^w = -mn \frac{\theta_s}{\left[\dfrac{1}{\ln\left(e + (\psi/a)^n\right)}\right]^{m+1}} \frac{\psi^{n-1}}{ea^n + \psi^n} \tag{10}$$

where: $e = 2.718$ (i.e., natural log base).

The transient water flow equation (Eq. 7) along with the equation of a soil-water characteristic curve (Eq. 8) and a permeability function (Eq. 9), can be used to predict pore-water pressure profiles (i.e., suction profiles) at different times during a seepage process. The suction profiles can then be used to compute the suction change for the stress-deformation analysis. The deformations due to changes in suction during any time period can then be predicted by specifying the initial and final soil suction profile.

For steady state seepage, only the coefficient of permeability is required because the time dependent term in Eq. (7) disappears and the storage function disappears. Gardner's equation(1958) is used for steady-state flow analyses.

$$k = \frac{k_s}{1 + a\left(\dfrac{\psi}{\rho_w g}\right)^n} \tag{11}$$

where: ψ = soil suction (kPa), k_s = coefficient of permeability at saturation, a = constant inversely proportional to the breaking point of the function, and n = constant related to the slope of the function.

The permeability function (Eq. 11) can be used to predict the pore-water pressure distribution (i.e., suction distribution) at equilibrium in the soil under specified boundary conditions. The suction profile then can be used to compute suction changes for the stress-deformation analysis to predict deformations.

Stress-Deformation

Equations of equilibrium for the soil structure of an unsaturated soil are:

$$\sigma_{ij,j} + b_i = 0 \tag{12}$$

where: σ_{ij} = components of the net total stress tensor, and b_i = components of the body force vector.

The partial differential equations for the soil structure can be derived from constitutive equations (Eq. 3) and the force equilibrium equations (Eq. 12). The governing partial differential equations in term of displacements in x- and y-direction (i.e., u and v) for plane strain loading ($d\varepsilon_z = 0$) are as follows:

$$\frac{\partial}{\partial x}\left\{c\left[(1-\mu)\frac{\partial u}{\partial x}+\mu\frac{\partial v}{\partial y}-\frac{(1+\mu)}{H}(u_a-u_w)\right]\right\}+\frac{\partial}{\partial y}\left\{G\left(\frac{\partial v}{\partial x}+\frac{\partial u}{\partial y}\right)\right\}=0 \quad (13)$$

$$\frac{\partial}{\partial x}\left\{G\left(\frac{\partial v}{\partial x}+\frac{\partial u}{\partial y}\right)\right\}+\frac{\partial}{\partial y}\left\{c\left[\mu\frac{\partial u}{\partial x}+(1-\mu)\frac{\partial v}{\partial y}-\frac{(1+\mu)}{H}(u_a-u_w)\right]\right\}+\rho g=0 \quad (14)$$

where: $c=\frac{E}{(1-2\mu)(1+\mu)}$, $G=\frac{E}{2(1+\mu)}$, ρ = density of the soil, and g = acceleration due to gravity.

Equations 13 and 14 are non-linear since the elastic moduli are functions of the stress state variables. The elastic moduli, E and H, can be calculated from the volume change indices, initial void ratio and Poisson's ratio as follows (Hung and Fredlund, 2000):

$$E=n_t(\sigma_{ave}-u_a)_{ave} \quad (15)$$

$$n_t=\frac{4.605(1+\mu)(1-2\mu)(1+e_0)}{C_t}$$

$$H=n_m(u_a-u_w)_{ave} \quad (16)$$

$$n_m=\frac{4.605(1+\mu)(1+e_0)}{C_m}$$

where: n_t = coefficient that relates net normal stress with elastic modulus E, n_m = coefficient that relate matric suction with elastic modulus H, $\sigma_{ave}=(\sigma_x+\sigma_y)/2$, $(s_{ave}-u_a)_{ave}$ = average of the initial and final net normal stress for an increment, and $(u_a-u_w)_{ave}$ = average of the initial and final matric suction for an increment.

Equations 13 and 14 can be used to compute the displacements in horizontal and vertical directions under an applied load or due to changes in matric suction. These equations can also be used to compute the induced stresses in the soil under an applied load. Because of soil property non-linearity, an incremental procedure is used to obtain the solution of these equations. In the incremental procedure, the values of elastic moduli E and H are assumed to be unchanged within each stress and strain increment, but are changed from one loading increment to another.

Example Problems

Three typical volume change problems that are often encountered in engineering practice are analyzed. The matric suction conditions in the soil mass were first predicted by performing a saturated/unsaturated water flow analysis. Then the deformations in soils caused by changing matric suction were predicted by performing a stress-deformation analysis. Both seepage and stress-deformation analyses are performed using a Partial Differential Equation Solver, PDEase2D.

Example 1: Influence of Trees to Surrounding Soil

The first example problem is associated with the deformation in soil near a line of trees. It is assumed that the trees are planted in a line at every 5 m. The example considers a 10 m thick layer of clay soil. The coefficient of permeability of the soil is described using Gardner's (1958) equation with a saturated coefficient of permeability equal to 5.79×10^{-8} m/s (i.e., 5 mm/day), parameters a and n equal to 0.001 and 2, respectively. The initial void ratio of the soil is equal to 1.0, and the volume change index with respect to matric suction, C_m is equal to 0.2.

The initial matric suction is taken to be hydrostatic with an unchanged ground water table at the 15 m depth. This represent the water content conditions in soils in the winter when water uptake by the trees is low. It is then assumed that one tree will extract 0.3 m^3 of water per day in summer and the steady state condition is attained. The value of 0.25 to 0.5 m^3/day was suggested by Perpich et. al. (1965). The water uptake zone for the trees is from the 1 m to 3 m depth, with uptake rate decreasing linearly with depth. This pattern of tree root water uptake was suggested by de Jong (2000).

Deformations in the soil profile due to water uptake by trees (from initial to final matric suction state) were predicted. The elastic modulus function (Fig. 3) with respect to matric suction for the soil was computed using a given initial void ratio, volume change index, and assumed Poisson's ratio equal to 0.3. The function can be calculated from Eq. 16 as follows:

$$H = 59.9(u_a - u_w)_{ave} \tag{17}$$

The initial and final matric suction conditions used to predict deformations in the stress-deformation analysis are obtained from the steady state seepage analysis

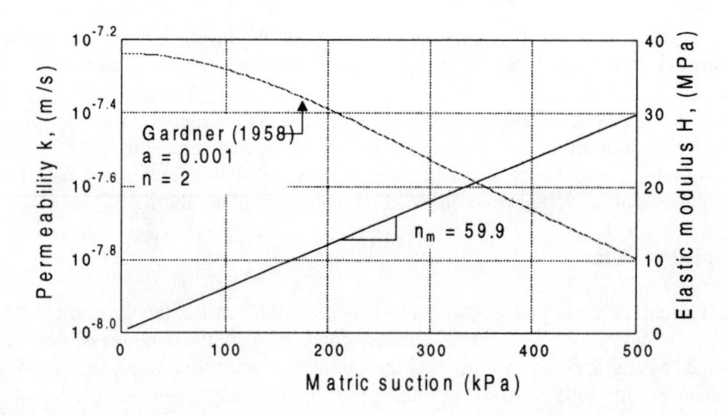

Figure 3. Permeability function, k and elastic modulus function, *H*, Example 1

using the coefficient of permeability function shown in Fig. 3. For boundary conditions for initial matric suction condition, a -15 m total head was specified at the lower boundary and a zero total head was specified at other boundaries. For final matric suction conditions, a -15 m total head was specified at the lower boundary, a boundary outflow value was specified along the left side of the soil domain at a depth from 1 m to 3 m and zero total head was specified at other boundaries. The outflow value decrease linearly from 15 mm/day at the 1 m depth to zero mm/day at the 3 m depth. This boundary condition represents 0.3 m^3/day of water being extracted by one tree from the soil (i.e., 2 sides x 2 m depth x 5 m wide x 15 mm/day = 0.3 m^3/day).

The matric suction distributions in the soil at equilibrium are shown in Figs. 4 and 5 for initial and final conditions, respectively. The initial matric suction varied from 147 kPa at ground surface to 49 kPa at 10 m depth. The final matric suction varied from 260 kPa at tree root to 49 kPa at 10 m depth. The matric suction change was a maximum at the tree root level and decreased with distance from the tree.

The deformations in the soil profile due to water uptake by tree roots was then predicted through the use of the stress deformation analysis. The boundary condition for the stress-deformation analysis involved having the soil free to move in the vertical direction and fixed in horizontal direction at the left and right sides of the domain. The lower boundary was fixed in both directions.

Figure 6 presents contours of vertical displacement in the soil for this

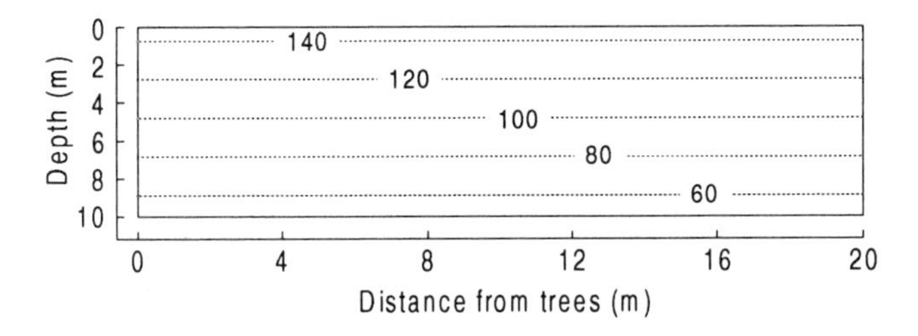

Figure 4. Initial matric suction (kPa) profile, Example 1

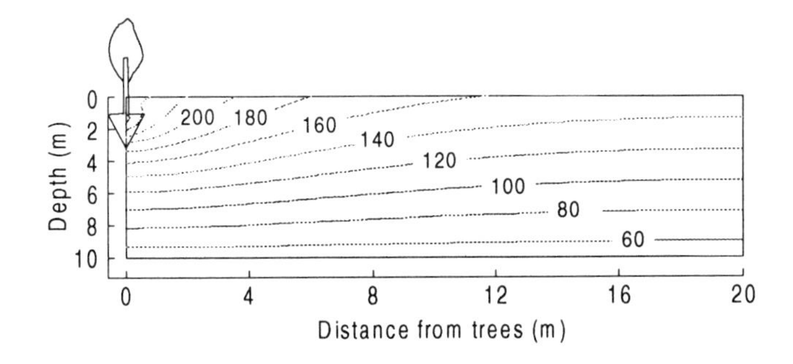

Figure 5. Final matric suction (kPa) profile, Example 1

example, with the volume change index with respect to matric suction equal to 0.2 and the tree-root uptake rate equal to 0.3 m^3/day per tree. The ground movements at various depths are shown in Fig. 7. It can be seen that the movements near the ground surface were quite large within a horizontal distance of about 4 m from the trees. The movements decreased rapidly with distance until at 12 m. The displacements also decreased rapidly with depth. The predicted displacements shown in Fig. 7 have the same pattern of displacements as these monitored by Bozozuk and Burn (1960). At ground surface, a value of settlement of 85 mm at tree location decreased to 40 mm at 8 m from the trees. At 4 m from the trees, a settlement of 65 mm at ground surface decreased to 40 mm at the 3 m depth, and about 20 mm at the 5 m depth.

The example was also analyzed for the case where the water uptake rate was 0.3 m^3/day and the volume change index varied with overburden pressure in soils (i.e., volume change index decreases linearly from 0.2 at ground surface to 0.05 at 10

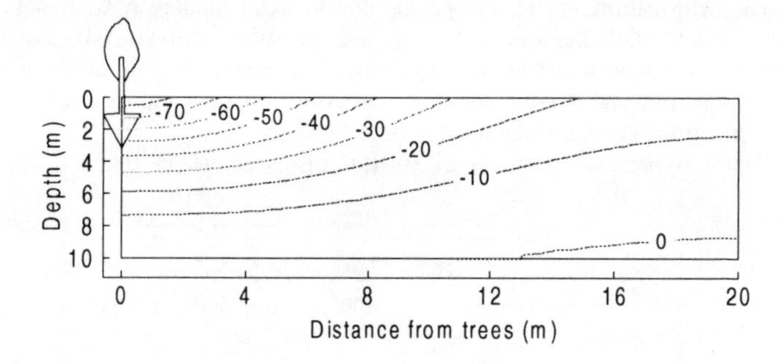

Figure 6. Contours of vertical displacement (mm), Example 1

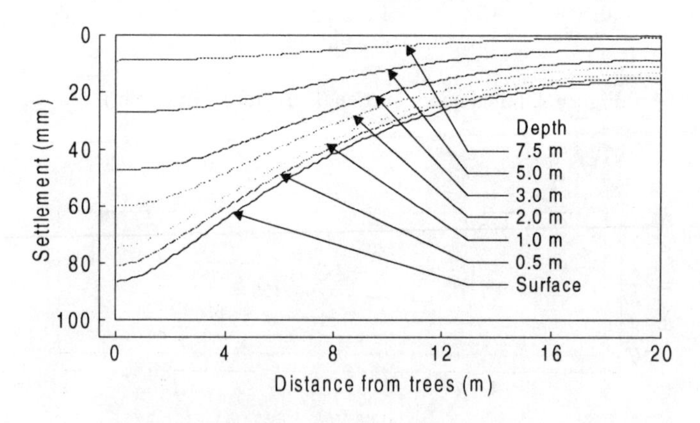

Figure 7. Variation with depth of ground movements near a line of trees, Example 1

m depth). Contours of vertical displacement are presented in Fig. 8 for this case. Most of displacements take place near ground surface.

Figure 9 presents contours of settlement for the case with a volume change index equal to 0.2 and a water uptake rate equal to 0.5 m^3/day. About 12% more settlement is obtained at tree location in comparison to the case when the water uptake rate was 0.3 m^3/day (i.e., (102 - 85)/85 = 12%).

Example 2: Influence of a Line of Trees to a House Footing

Example 2 simulates the settlement of a house induced by a line of trees growing close to the house. This example is illustrated in Fig. 10. The house foundation is placed at a 2 m depth in a clay soil layer that is 10 m thick. The line of trees is 4 m from the house. Water uptake rate of the trees is 0.5 m^3/day per tree and the water uptake zone is from 1 m to 3 m depth. It is assumed that the ground water table is unchanged at 15 m below ground surface. A layer of 0.3 m thick of concrete with elastic modulus of 100,000 kPa is used to describe the foundation and basement walls of the house. Soil properties are the same as those used in Example 1 (Fig. 3).

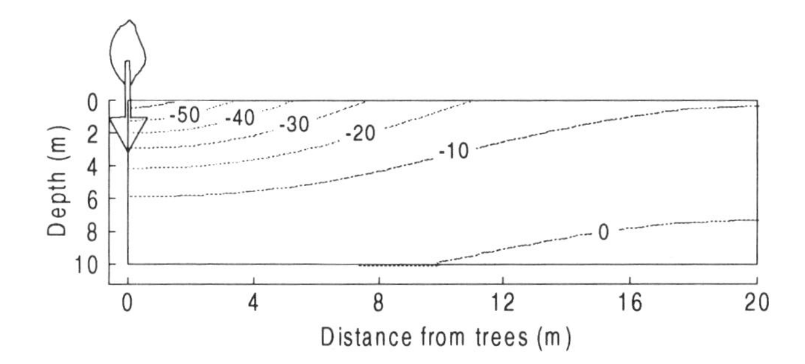

Figure 8. Contours of vertical displacement (mm) when volume change index varied from 0.2 at surface to 0.05 at the 10 m depth, Example 1

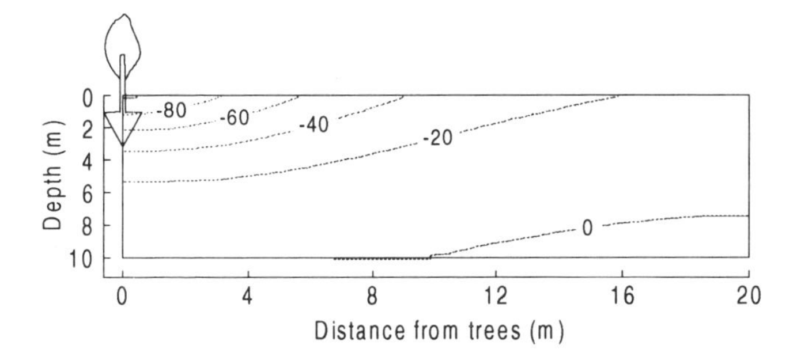

Figure 9. Contours of vertical displacement (mm) when water uptake rate was 0.5 m^3/day and volume change index equaled to 0.2, Example 1

Matric suction conditions in the soil profile were obtained through steady state unsaturated seepage analyses. The initial matric suction profile is the same as that shown in Fig. 4. The final matric suction profile is shown in Fig. 11. Final matric suction varied from 409 kPa at tree root to 49 kPa at lower boundary of the soil domain. The contours of changes in matric suction are presented in Fig. 12. The closer to the tree, the more change in suction is observed. Figure 12 also showed a similar pattern of moisture deficit near trees presented in Biddle (1983). The results of stress-deformation analysis are shown in Figs. 13 and 14 as contours of horizontal displacement and contours of vertical displacement. A maximum foundation settlement of 80 mm and minimum settlement of 25 mm was observed. A maximum settlement in the soil profile took place at tree location and decreased with horizontal distance and depth.

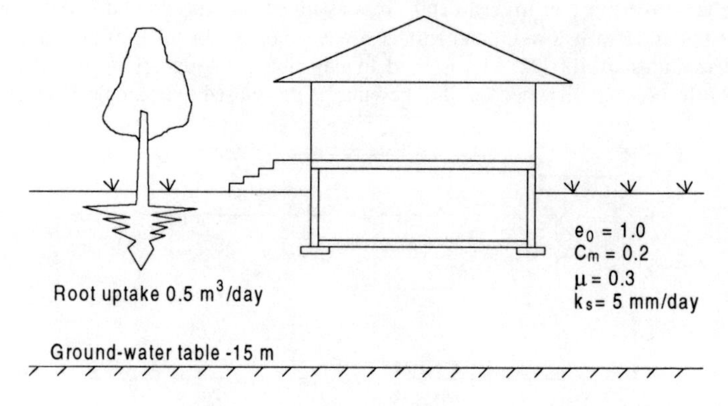

Figure 10. Illustration of Example 2

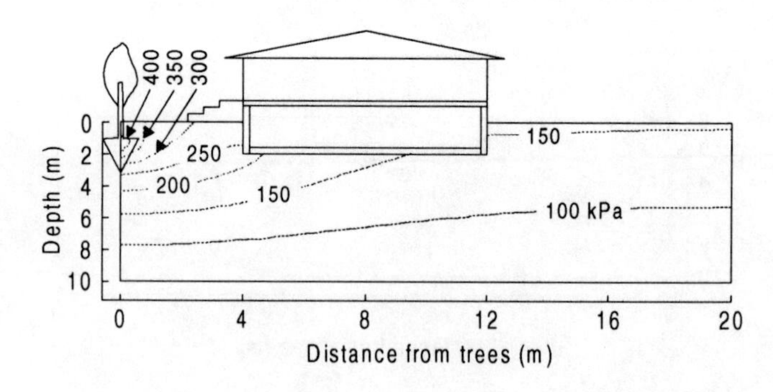

Figure 11. Contours of final matric suction (kPa), Example 2

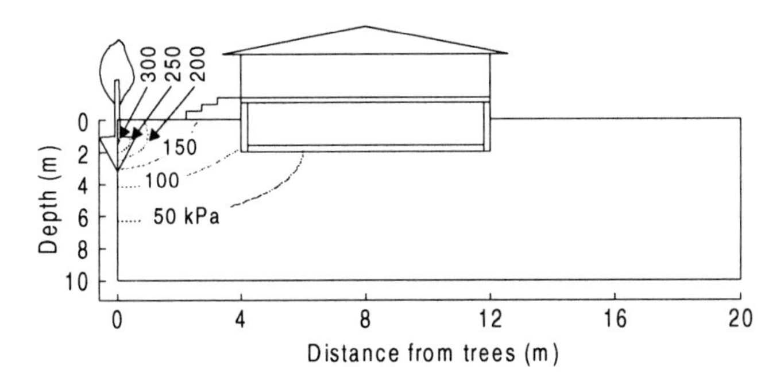

Figure 12. Contours of changes in matric suction (kPa), Example 2

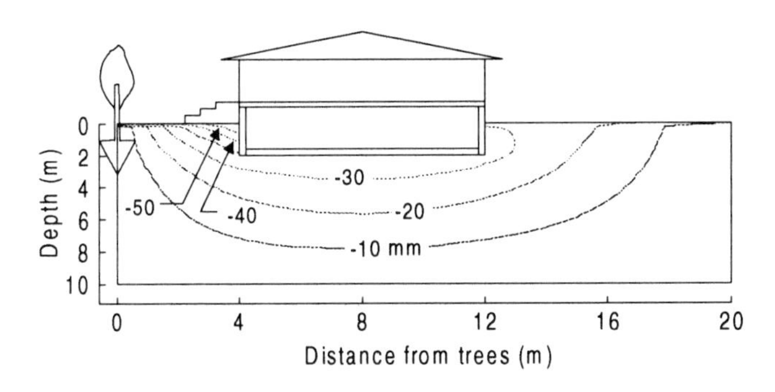

Figure 13. Contours of horizontal displacement (mm), Example 2

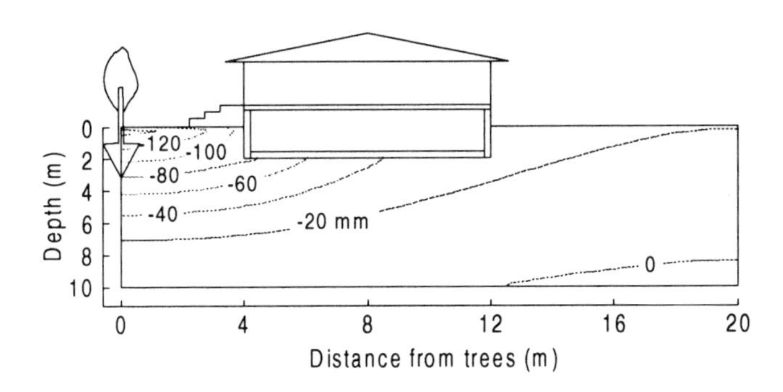

Figure 14. Contours of vertical displacement (mm), Example 2

Example 3: Influence of Ground Surface Flux to a Concrete Floor Slab

This example is presented to show a comprehensive volume change problem in an unsaturated soil. This example considers the deformations in a soil profile under a concrete floor slab due to applied load and changes in matric suction (i.e., water infiltration into the soil profile from surface). The deformations will be predicted for various predetermined elapsed times. The problem to be analyzed is illustrated in Fig. 15. A concrete slab, 0.3 m thick and 8 m wide is placed on a 5 m thick layer of swelling clay. All the soil properties to be used in the analysis are presented in Figs. 15 and 16. The soil has an initial suction of 700 kPa throughout the profile. The applied load on the slab, including its weight is 10 kPa. The right side of the slab is covered with an impermeable flexible layer. The grass at the left side is watered with the flux of 5.79×10^{-9} m/s.

The soil-water characteristic curve is described using the Fredlund and Xing (1994) equation, with the parameters *a* equal to 100 kPa, *n* equals 1.5 and *m* equals 1. The permeability function is described using the equation proposed by Leong and Rahardjo (1997) based on the Fredlund and Xing (1994) function for the soil-water characteristic curve with *p* equal to 1. Because the parameter *p* is equal to 1, the soil-water characteristic curve and the permeability function have the same shape and are shown in Fig. 16. The concrete slab has elastic modulus with respect to net normal stress equal to 1 MPa.

The elastic modulus function with respect to net normal stress, E, can be calculated from the given compressive index with respect to net normal stress, initial void ratio and assumed Poisson's ratio using Eq. 15 and written as follows:

$$E = 12.89(\sigma_{ave} - u_a)_{ave} \tag{18}$$

The elastic modulus function with respect to matric suction, *H*, is calculated using Eq. 16 for the soil with given initial void ratio, swelling index with respect to matric suction, and assumed Poisson's ratio. This function can be written as follows:

$$H = 184.20(u_a - u_w)_{ave} \tag{19}$$

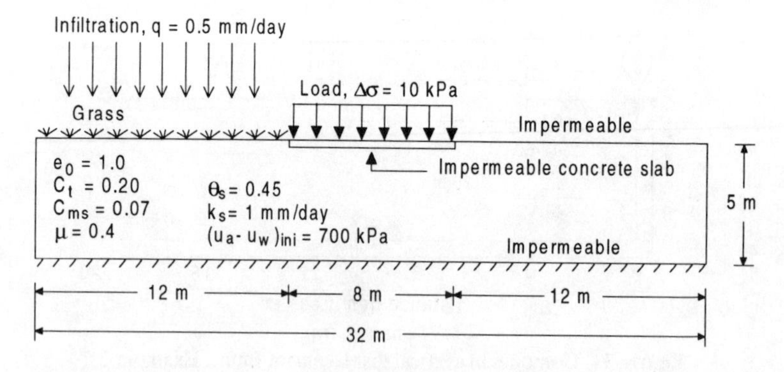

Figure 15. Illustration of Example 3

The matric suction conditions in the soil at various elapsed times can be predicted by performing an unsaturated transient seepage analysis. For the transient seepage analysis, a boundary flow value of 5.79×10^{-9} m/s is specified along the uncovered surface and zero flux is specified at the other boundary sides. The initial matric suction is equal to 700 kPa. Matric suction profile at 400 days of constant infiltration is presented in Fig. 17 as a typical matric suction distribution in soils.

Deformation in the soil mass due to loading can be assumed to respond immediately, while the deformation due to wetting is a time dependent process. Therefore, the stress-deformations due to loading and due to wetting need to be analyzed independently. A stress-strain analysis is performed in six stages. One stage is for changes in net normal stress and the other five are for changes in suction. In the first stage, the deformation due to loading (i.e., change in net normal stress) is modeled. The second stage is carried out to determine the deformation due to changes in matric suction after 100 days of watering the grass. In this stage, the initial matric suction profile has a constant value of 700 kPa, and the final matric suction profile is computed from the water flow analysis, at the first elapsed time of 100 days. The third stage gives the deformation due to changes in suction in the next 100 days (i.e., from 100th to 200th day). The fourth, fifth and sixth stages are for the

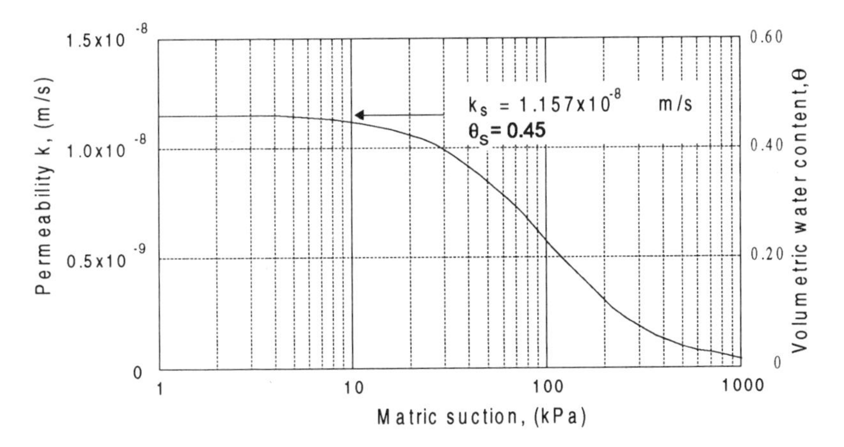

Figure 16. Permeability function and SWCC, Example 3

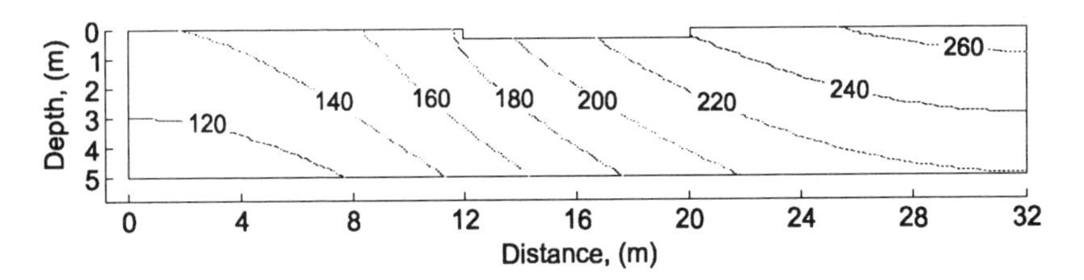

Figure 17. Matric suction (kPa) profile at day 400, Example 3

time periods from 200th to 300th day, 300th to 400th day, and 400th to 450th day, respectively.

The boundary conditions are specified for the stress-deformation analysis as follows. At the left and right sides of the domain, the soil is free to move in the vertical direction while it is fixed in the horizontal direction. The lower boundary is fixed in both directions.

The initial vertical total stress is calculated from total stress theory and the initial horizontal stress is calculated using a coefficient of earth pressure at rest, K_O, equal to 0.67.

The vertical displacement and horizontal displacement due to wetting for each stage of the analysis are predicted. The cumulative vertical displacements at the surface are presented in Fig. 18 for each stage. Figures showing distribution of horizontal displacements are not presented due to lack of space in the paper. The maximum calculated heave increases from 20 mm during the first 100 days to 45 mm in the stage from "Day 300" to "Day 400" of infiltration, and in the last stage (i.e., in the last 50 days), a maximum heave of 46 mm was predicted. Small horizontal displacements were observed during the infiltration (i.e., maximum of 1.5 mm for the first 100 days and 5 mm for the last 50 days).

Figure 19 presents the cumulative vertical displacements below the concrete slab. The vertical displacements caused by loading are presented by the contours in Fig. 20. A maximum settlement of 65 mm was predicted for the concrete slab.

Total heave due to wetting for 450 days is presented in Fig. 21. As much as 155 mm of heave was predicted at the surface of the exposed area (grass portion), and 90 mm of heave was predicted at the covered portion. The differential heave of the concrete slab was about 30 mm. The horizontal displacements take place only in the grassed portion, where 10 mm of horizontal displacements was predicted. The

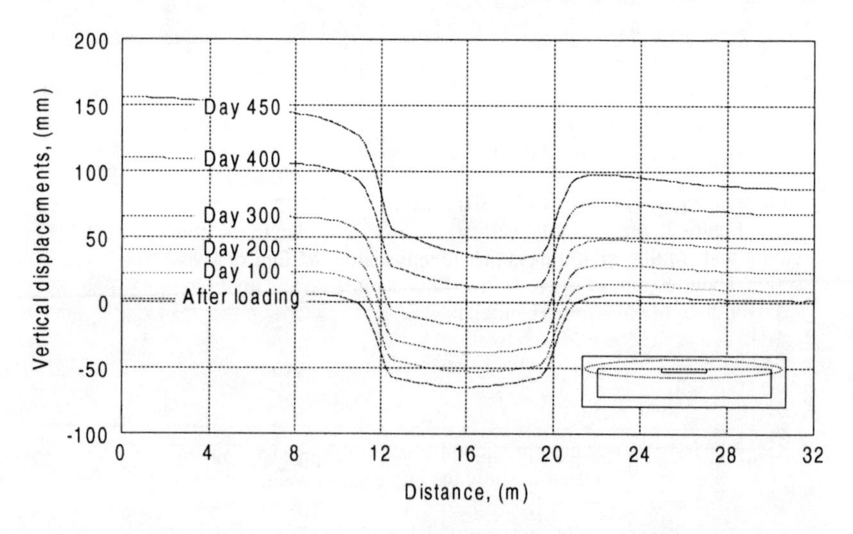

Figure 18. Cumulative vertical displacement at surface at various times, Example 3

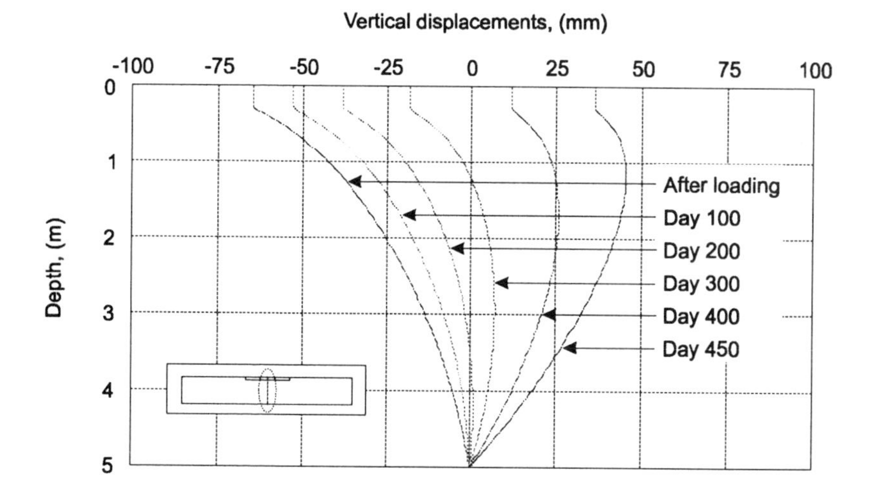

Figure 19. Cumulative vertical displacements versus depth below the slab at various times, Example 3

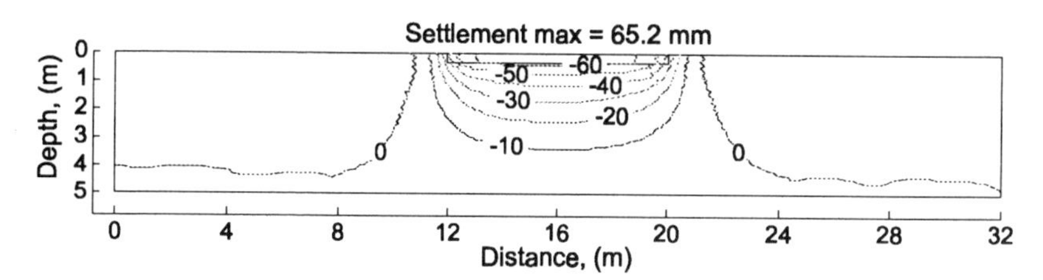

Figure 20. Contours of vertical displacement due to loading (mm), Example 3

differential heaves predicted depend on the rate of infiltration; a lower rates of infiltration results in lower differential heave.

Figure 22 presents total vertical displacements due to the combined effects of loading and wetting in 450 days. There was about 155 mm of heave in the grassed portion, about 40 mm of heave below the concrete slab and about 90 mm of heave in the cover area. Maximum horizontal displacements of 20 mm took place at about 1.5 m depth below the end of the concrete slab.

Conclusion

A method of volume change prediction based on the general theory of unsaturated soil behavior have been used to study the influence of water uptake by tree roots and water infiltration to surrounding soils and nearby structures. Results of the analyses appear to be reasonable and consistent with measured values. The results show that magnitude of deformation of structures and soils close to the trees decreased with vertical and horizontal distance from trees, and with decreasing of water uptake rates. It is possible to combine unsaturated soil theory for seepage and stress analysis to study the influence of vegetation and environmental conditions on light engineered structures.

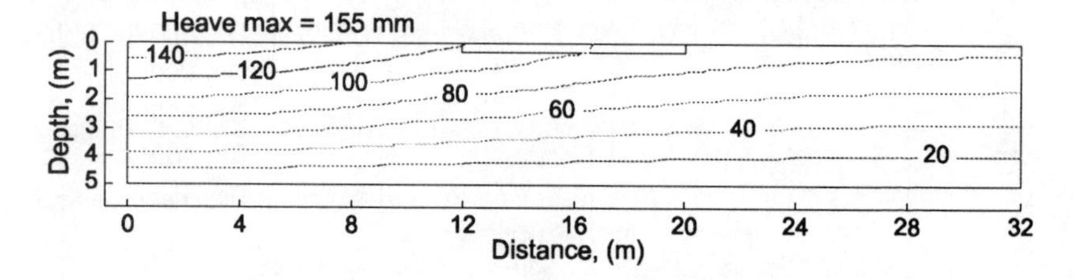

Figure 21. Contours of vertical displacement due to wetting after 450 days (mm), Example 3

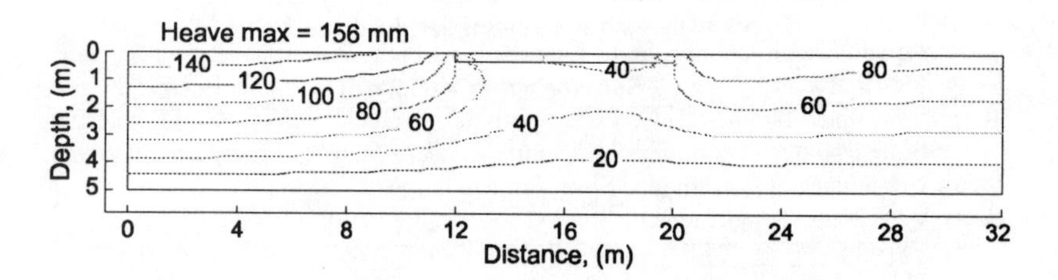

Figure 22. Contours of vertical displacement due to loading and wetting after 450 days (mm), Example 3

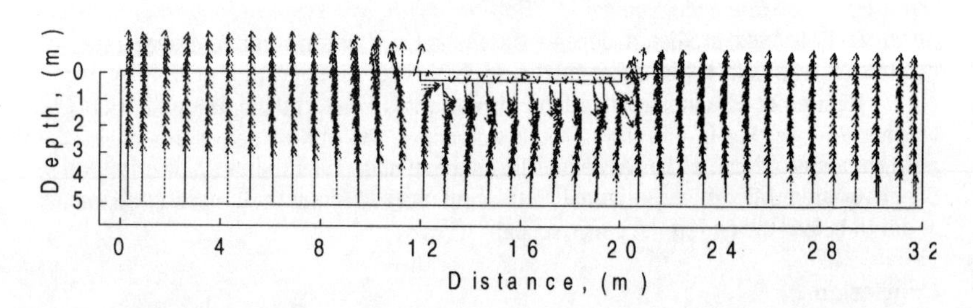

Figure 23. Distribution of deformation vectors due to loading and wetting after 450 days (mm), Example 3

Reference

Biddle, P.G. 1983. Patterns of soil drying and moisture deficit in the vicinity of trees on clay soils. Geotechnique: The Influence of Vegetation on the Swelling and Shrinking of Clays, **33**(2): 107-126.

Bozozuk, M. and Burn, K.N. 1960. Vertical Ground Movements Near Elm Trees. Geotechnique, **10**: 19-32.

de Jong, E. 2000. Personal communication. College of Agriculture, University of Saskatchewan.

Fredlund D.G. 1995. The Scope of Unsaturated Soils Mechanics: An Overview. Proceedings, 1st International Conference on Unsaturated Soils, Paris, France, Vol. 3, pp. 1155-1177.

Fredlund, D.G. and Xing, A. 1994. Equation for the Soil-Water Characteristic Curve. Canadian Geotechnical Journal, **31**(3): 521-532.

Fredlund, D.G. and Rahardjo, H. 1993. Soil Mechanics for Unsaturated Soils. John Wiley & Sons, New York, 560 p.

Fredlund, D.G., and Morgenstern, N.R. 1977. Stress State Variable for Unsaturated Soils. Journal of Geotechnical Engineering Division, Proceedings, American Society of Civil Engineering (GT5), **103**: 447-466.

Gardner, W.R. 1958. Some Steady State Solutions of the Unsaturated Moisture Flow Equation with Application to Evaporation from a Water Table. Soil Science, **85**(4): 228-232.

Hung, V.Q. and Fredlund, D.G. 2000. Volume Change Predictions in Expansive Soils Using a Two-dimensional Finite Element Method. Proceeding of Asian Conference on Unsaturated Soils, Singapore, pp. 231-236.

Hung, V.Q. 2000. Finite Element Method for the Prediction of Volume Change in Expansive Soils. M.Sc. Thesis, University of Saskatchewan, Saskatoon, SK, Canada, 314 p.

Leong, E.C. and Rahardjo, H. 1997. Permeability Functions for Unsaturated Soils. Journal of Geotechnical and Geoenvironmental Engineering, ASCE, pp: 1118-1126.

PDEase2D 3.0 Reference Manual, 3rd Edition. 1996. Macsyma Inc., Arlington, MA, 02174 USA.

Perpich, W.M., Lukas, R.G., and Baker, Jr. C.N. 1965. Desiccation of Soil by Trees Related to Foundation Settlement. Canadian Geotechnical Journal, **1**(2): 23-39.

van Genuchten, M.T. 1980. A Closed-formed Equation for Predicting the Hydraulic Conductivity of Unsaturated Soils. Soil Science Society of America Journal, **44**: 892-898.

Estimating Soil Swelling Behavior Using Soil Classification Properties

Andrew P. Covar[1], M. ASCE and Robert L. Lytton[2], F. ASCE

Abstract

There exists a need for a method of estimating soil swelling behavior for the design and practitioner community. Desirable method attributes include; low cost, reasonable accuracy and technical soundness. The method should be usable for a wide range of soil types. Practitioners should be able to use local soils testing services to get the data required by the method. The NRCS Soil Survey Laboratory (SSL) soils database is a large, quality controlled resource that provides the data for the development of such a method. The presented method allows for the estimation of the suction compression index using Atterberg limits, particle size classification, and the coefficient of linear extensibility (COLE) values as contained in the SSL database. Using a 6500 sample subset of the SSL database, a series of charts are presented yielding suction compression index values for various mineralogically based groups. The proposed method provides quick and stable prediction of an important soil property using low cost and commonly available soil test procedures.

Introduction

There exists a need for a method that provides geotechnical practitioners with an estimate of soil swell characteristics as a part of the design process for building slab and mat foundations, piers and other support structures. Desirable attributes of such a methods include; low cost, straightforward to use, technically supportable, and reasonably accurate. In addition the method should provide results for a broad

---

[1]Doctoral Candidate, Texas A&M University, Mail Stop 3136, College Station, TX 77843-3136; phone 512-996-8070; acovar@tamu.edu.
[2]F.J. Benson Chair Professor of Civil Engineering, Texas A&M University, College Station, TX 77843-3136, phone 979-845-8211, fax 979-845-0278, rlytton@tamu.edu.

geographical coverage and provide answers using analyses within the capability of most conventional geotechnical laboratories.

Presented herein are the results of an examination of the use of Atterberg limits, particle size information, and cation exchange activity to produce such a predictive method.

Earlier Predictive Methods

The use of Atterberg limits as predictors of soil behavior has been common since their development. The Testing is relatively inexpensive, re-producible, and fast compared to many other tests. Given this long history, there is wide geographic coverage for test results.

Holtz and Gibbs (1956) developed a soil swell classification shown in Table 1 using index tests. Pearring (1963) used cation exchange capacity, CEC, and plasticity as two parameters to classify soils as to a predominant mineral type. Pearring normalized these two parameters based on the percent fine clay content. This normalization yielded two new parameters, the activity ratio (Ac) and the cation exchange activity (CEAc) as follows;

$$Ac = \frac{PI\%}{\frac{(\% - 2\text{micron})}{(\% - \text{No.200 sieve})} x100}$$

$$CEAc = \frac{CEC \frac{\text{millieq.}}{\text{100gm of dry soil}}}{\frac{(\% - 2\text{micron})}{(\% - \text{No.200sieve})} \text{x } 100}$$

Figure 1 illustrates the classification developed by Pearring (1963).

Seed et. al. (1962) developed a chart based on Ac and the percent clay fraction. The chart is shown as Fig. 2. Snethen et al. (1977) re-evaluated criteria for predicting soil swell and found that the soils' liquid limit, plasticity index and soil suction at natural moisture content were the best indicators of potential swell. The resulting classification system is shown in Table 2. Lytton (1977, 1994) presented an expression relating the volumetric change in a soil sample due to changes in water potential. The relationship includes both water potential and stress terms as follows;

$$\frac{\Delta V}{V} = -\gamma_h \log_{10}\left(\frac{h_f}{h_i}\right) - \gamma_\sigma \log_{10}\left(\frac{\sigma_f}{\sigma_i}\right)$$

where;

h_i, h_f are initial and final water potential,
N_i, N_f are normal stress terms, and
3_h and 3_N are the matric suction compression index and the mean principal stress compression index respectively.

McKeen (1981) used a mineralogical classification similar to that of Pearring (1963), defining regions charted against Ac and CEAc axes. For each of the identified zones, a value for the corresponding suction compression index is given after being adjusted to a 100% fine clay fraction. The actual suction compression index for the soil portion finer than the No. 200 sieve is proportional to the 100% fine clay index by the following equation;

$$\gamma_h = \gamma_{100}\left[\frac{\% - 2\text{ micron}}{\% - \text{No.200 sieve}}\right]$$

The COLE test represents the fractional change in a clod sample resulting from changes in moisture content. The classification chart, including predicted COLE values, is shown in Fig. 3. Hamberg (1985) updated the classification chart and included an adjustment for clay percentage as shown in Fig. 4. The approach continued to be refined in Nelson and Miller (1992) that produced a more simple general classification scheme using CEAc and Ac axes as shown in Fig. 5.

The NRCS Database

The United States Department of Agriculture, Natural Resources Conservation Service has created a national database of soil samples and test results. At the present time there are data for more than 25,000 soil pedons from all 50 states, other U.S. territories and several foreign countries. The stated objective of the program is to obtain representative samples of soils throughout the United States and its territories to provide a consistent analytical view of these soils' chemical, mineralogical, and physical characteristics. Each soil sample was collected by experienced soil scientists at each horizon for future analyses. Different methods of sampling were used depending on the specific tests to be used. Clod samples were collected at each horizon for COLE determinations, for example.

Data from these analyses are compiled by the Soil Survey Laboratory (SSL) of the National Soil Survey Center. Most of the data in the present database were obtained over the last 40 years with approximately 75% of the data being obtained in the last 25 years. The SSL database may be accessed online at http://www.statlab.iastate.edu/soils/ssl. The database is also available on CD-ROM from the SSL. The CD-ROM version is in the form of a relational database approximately 200 MB in size containing the results of analyses on approximately 130,000 soil samples.

Data Analysis

The SSL database as selected as the basis for a re-examination of soil swell behavior as it is affected by certain index properties. SSL database features that

Table 1. Expansive Soil Classification from Holtz and Gibbs (1956)

Colloid Content (% minus .0001 mm)	Plasticity Index	Shrinkage Limit	Probable Expansion (%Vol)	Degree of Expansion
>28	>35	<11	>30	Very high
20-31	25-41	7-12	20-30	High
13-23	15-28	10-16	10-20	Medium
<15	<18	>15	<10	Low

Table 2. Expansive Soil Classification based on Atterberg Limits from Snethen et al. (1977)

LL (%)	PI (%)	Natural Soil Suction	Potential Swell %	Potential Swell Classification
>60	>35	>4	>1.5	High
50-60	25-35	1.5-4	0.5-1.5	Marginal
<50	<25	<1.5	<0.5	Low

were important include; wide geographic coverage, the quantity of analyses available and extensive quality control in sampling, analysis and reporting. For this study, the SSL database was filtered to retain only those records that contained non-null results for the following tests;

Liquid limit
Plasticity index
Plastic limit
Cation exchange capacity
Coefficient of linear extensibility
% passing 2 micron
% passing No. 200 sieve

This data filtering produced a subset of the data containing approximately 6400 records. These data are shown in Fig. 6. The figure illustrates the broad distribution of the data. Next, the data records were partitioned according to Fig. 7 based on Casagrande (1948) and the Holtz and Kovacs (1981) mineral classification chart.

This partitioning step resulted in eight separate data groups, each representing a group with some mineralogical similarity. For each record a matric suction index was calculated according to the following expressions,

$$\gamma_h = \left(\frac{\gamma(\text{swelling case}) + \gamma(\text{shrinking case})}{2} \right)$$

$$\gamma(\text{swelling case}) = \left[\left(\frac{\text{COLE}}{100}+1\right)^3 - 1\right]$$

$$\gamma(\text{shrinking case}) = \left[1 - \frac{1}{\left(\frac{\text{COLE}}{100}+1\right)^3}\right]$$

The calculated (average) suction compression index was then adjusted to a 100% fine clay content.

Data for each of the eight separate mineralogical groups was then plotted as contoured surfaces on Ac-LL/%fine clay axes as shown in Fig. 8 through 15. These contoured data were created using a kriging algorithm per Cressie (1990). No explicit smoothing interpolation was used in creating the plotted surfaces. Table 3 provides summary statistics for each of the eight data groups. More simple versions of these surfaces have been prepared for Groups I though IV, TBPE (2000) and are shown as Fig. 16a to 16d.

Conclusions

The SSL database provides a rich source of analytical data for researchers and practitioners alike. The dataset may well be one of the largest available for the examination of wide ranging soil characteristics.

The method developed herein represents a refinement of earlier methods. The method builds on these earlier methods in that it is consistent in the use of low cost and easily available testing methods (Atterberg limits and soil particle size distributions) to predict soils properties and behavior. The method is "stable" in the sense that each mineralogical zone or group is explicitly defined, thus no arbitrary distinctions can affect the results. Within each group, the practitioner needs only the liquid limit, plasticity index and the fine clay fraction (%) to get an estimate of the suction compression index. The suction compression index can then be
explicitly to calculate shrink and swell behavior using previously published methods as Lytton (1977, 1994). Small changes in soil index properties result in small changes in the derived suction compression index within each mineralogically based group. The proposed method, therefore, provides quick and stable prediction of an important soil property using low cost and commonly available soil test procedures.

Table 3. Summary Statistics for Suction Compression Indices; Minimum, Maximum and Percentile Values

Zone	1	2	3	4	5	6	7	8
minimum	0.00	0.00	0.00	0.00	0.01	0.01	0.01	0.00
25%	0.05	0.04	0.03	0.03	0.03	0.02	0.04	0.03
50%	0.09	0.08	0.07	0.06	0.05	0.05	0.08	0.07
75%	0.14	0.12	0.12	0.10	0.09	0.07	0.14	0.14
95%	0.21	0.20	0.20	0.20	0.19	0.17	0.32	0.31
maximum	0.44	0.46	0.48	0.44	0.48	0.48	0.49	0.47
No. of Data Pts.	523	1328	2534	991	266	166	266	302

References

1. Casagrande, A. (1948). "Classification and Identification of Soils," Trans. ASCE, Vol. 30, No. 2, 211-213.

2. Cressie, N. A. C. (1990). "The Origins of Kriging," Mathematical Geology, Vol. 22, 239-252.

3. Holtz, W.G. and Gibbs, H.J. (1956). "Engineering Properties of Expansive Clays," Trans. of ASCE 121:641-677.

4. Holtz, R.D. and Kovacs, W.D. (1981). *An Introduction to Geotechnical Engineering*, Prentice-Hall, Englewood, NJ.

5. Lytton, R.L. (1977). "Engineering Properties of Expansive Soils," Presentation to the American Geophysics Union, Conference, San Francisco.

6. Lytton, R.L. (1990). "Prediction of Movement in Expansive Soils," ASCE Special Publication 40.

7. McKeen, R.G. (1981). "Design of Airport Pavement on Expansive Soils," Dept. of Trans., Federal Aviation Admin., Rept. No. DOT/FAA/RD-81/25.

8. Mckeen, R.G. and Hamberg, D. J. (1981). Characterization of Expansive Soils. Trans. Res. Rec. 790, Trans. Res. Board 73-78.

9. Nelson, J.D. and Miller, D.J. (1992). *Expansive Soils*, John Wiley & Sons, New York.

10. Pearring, J.R. (1963), "A Study of Basic Mineralogical, Physico-chemical and Engineering Index Properties of Laterite Soils," dissertation, Texas A&M University, College Station.

11. Seed, H.B., Mitchell, J.K. and Chan, C.K. (1962). "Studies of Swell and Swell Pressure Characteristics of Compacted Clays," Highway Res. Board Bulletin. 331:12-39.

12. Snethern, D. R. , Johnson, L. D. and Patrick, D. M. (1977). "An Evaluation of Expedient Methodology for Identification of Potentially Expansive Soils." Soils and Pavements Lab., U.S. Army eng. Experiment Sta., Vicksburg, MS, Rep. No. FHWA-RE-77-94.

13. TBPE (2000), Residential Foundation Design Subcommittee, Advisory Report, Texas Board of Professional Engineers, Austin, TX.

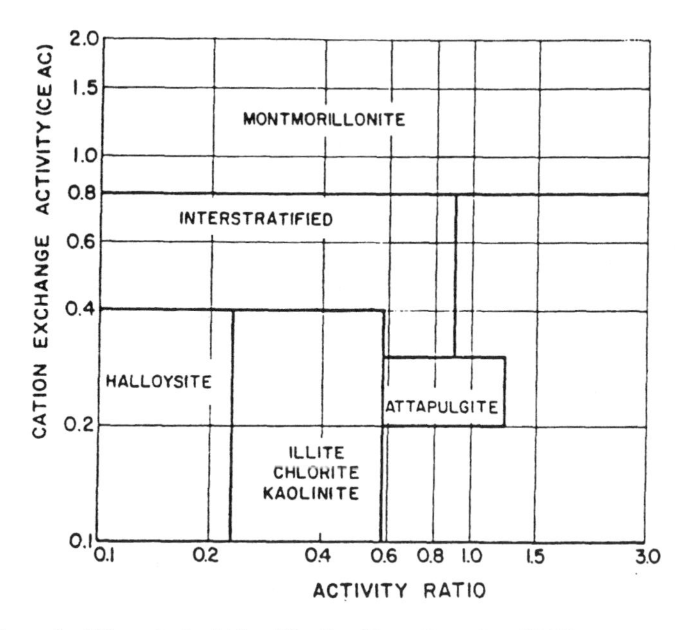

Figure 1. Mineralogical Classification From Pearring (1963)

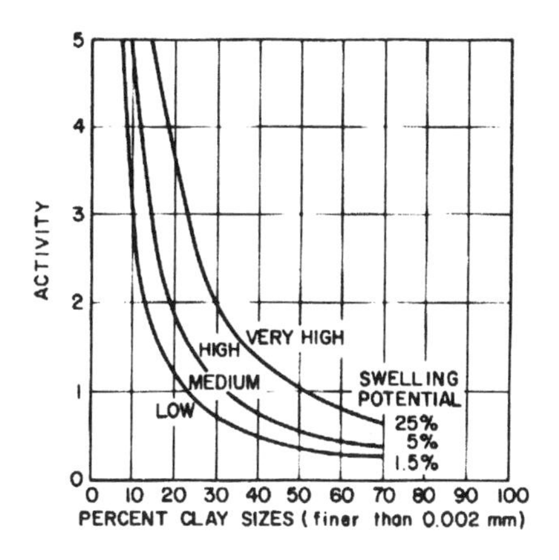

Figure 2. Soil Swell Potential Based on Size Fraction and Activity from Seed (1962)

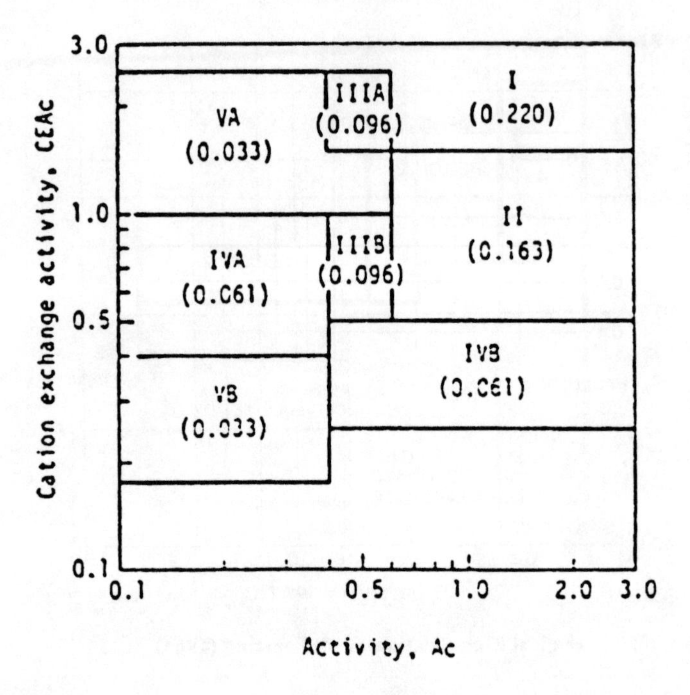

Figure 3. Classification Chart for COLE Values from McKeen and Hamberg (1981)

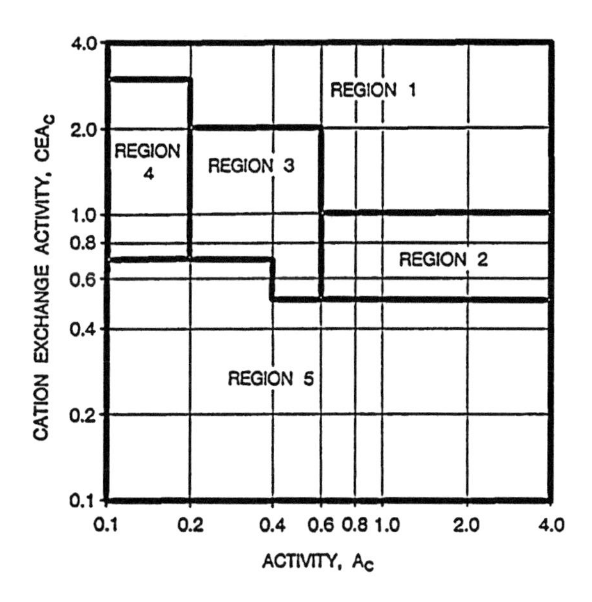

Region	A	B	n	r^2
1	0.23	-1.46	20	0.79
2	0.20	-1.14	45	0.77
3	0.14	-0.71	93	0.72
4	0.13	-0.31	21	0.51
5	0.09	-0.96	20	0.87

Figure 4. COLE Values bAsed on CEAc, Ac and Fine Clay from Hamberg (1985)

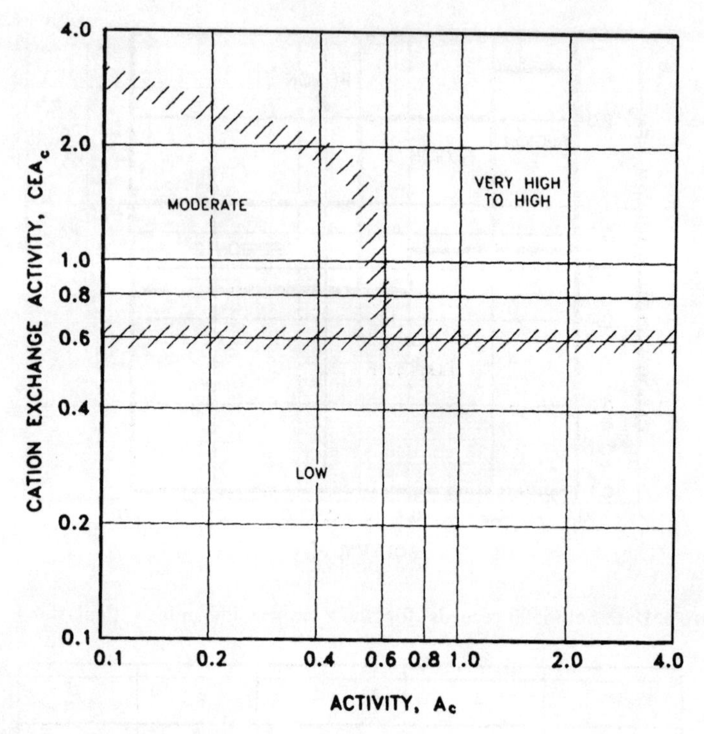

Figure 5. Expansion Potential as a function of CEAc and Ac from Nelson (1992)

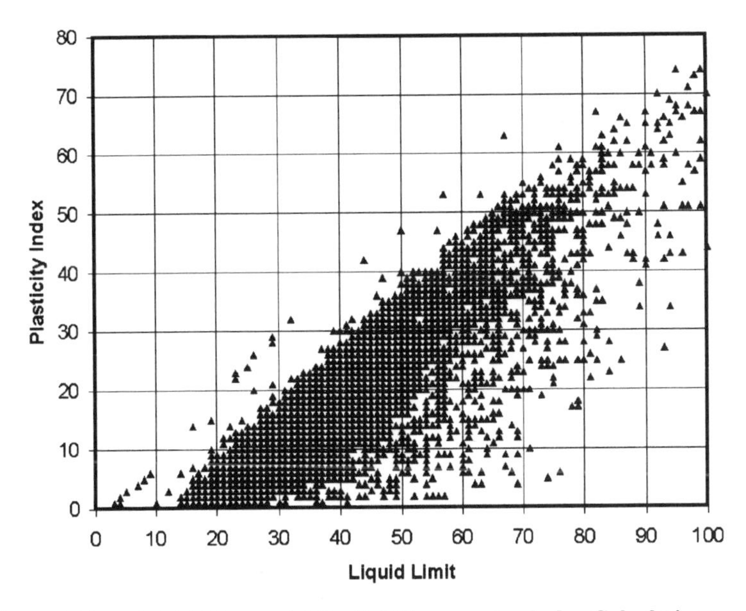

Figure 6. Data Set (6500 records) for Soil Compression Index Calculations

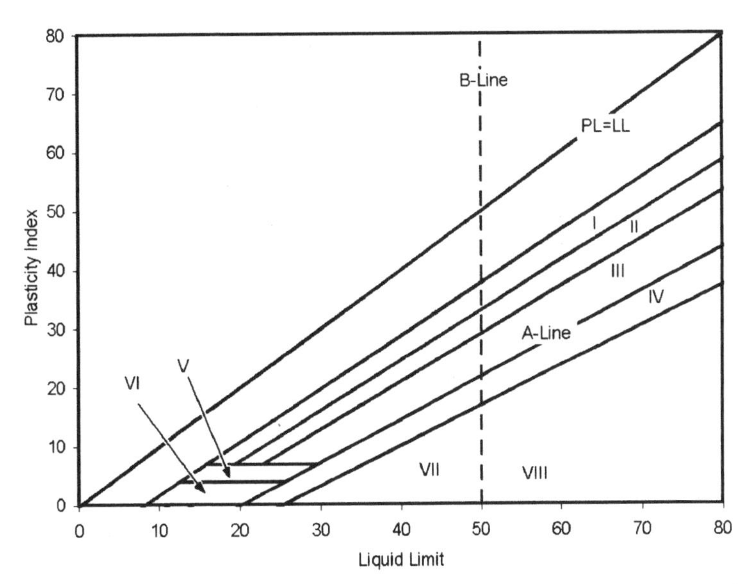

Figure 7. Data Filter for Partioning Data base on Mineralogical Types after Casagrande (1948) and Holz and Kovac (1981)

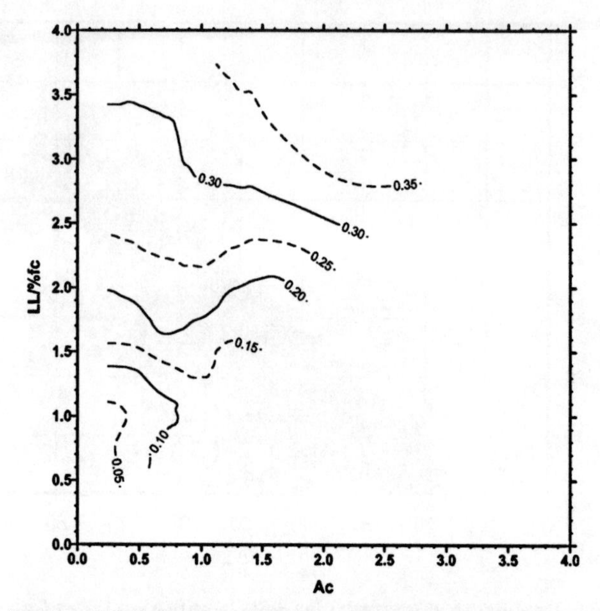

Figure 8. Predicted Soil Compression Index Values for Zone I

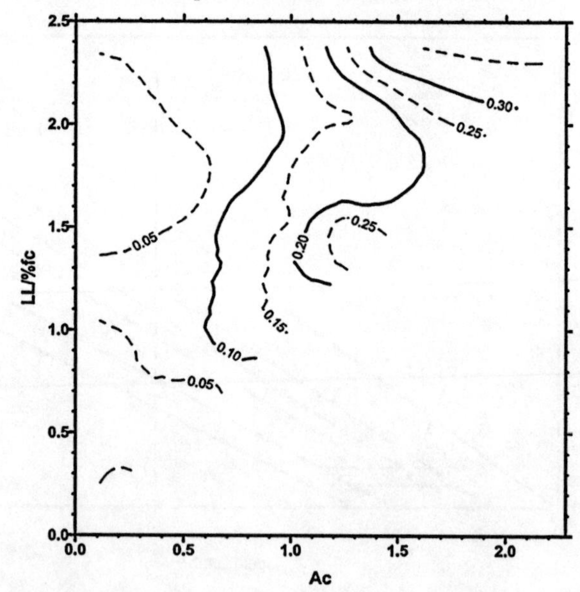

Figure 9. Predicted Soil Compression Index Values for Zone II

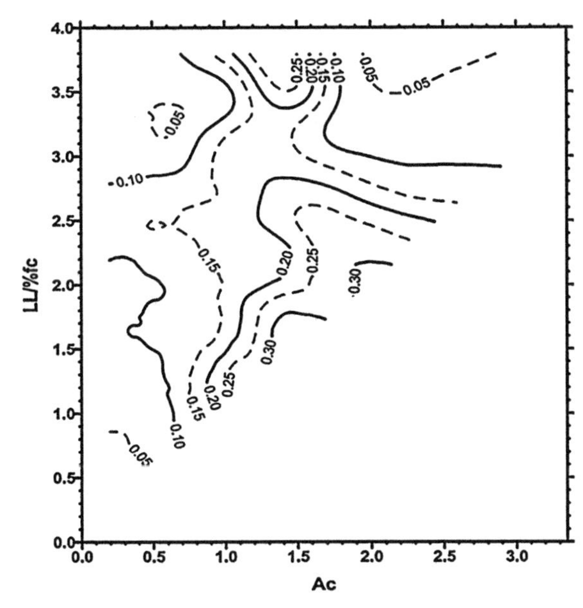

Figure 10. Predicted Soil Compression Index Values for Zone III

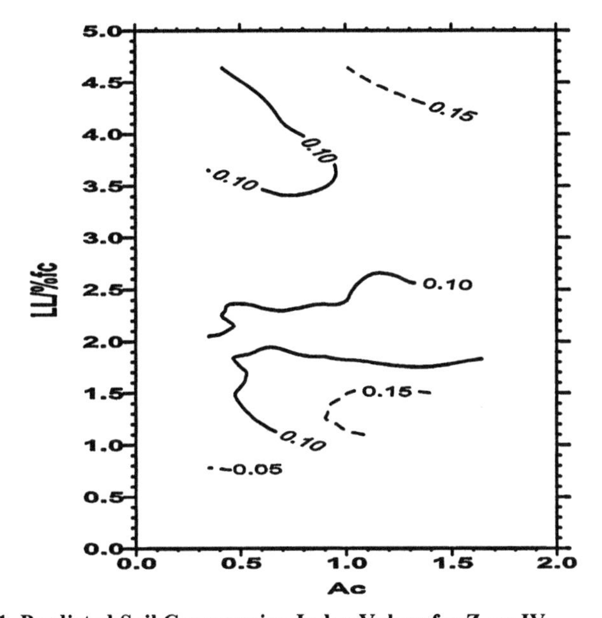

Figure 11. Predicted Soil Compression Index Values for Zone IV

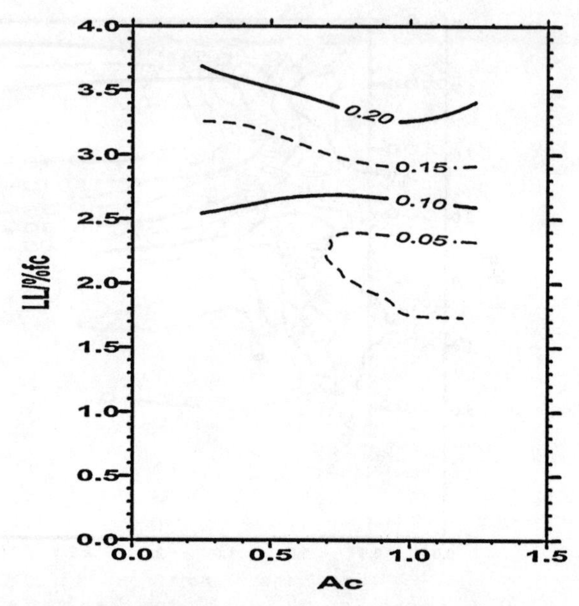

Figure 12. Predicted Soil Compression Index Values for Zone V

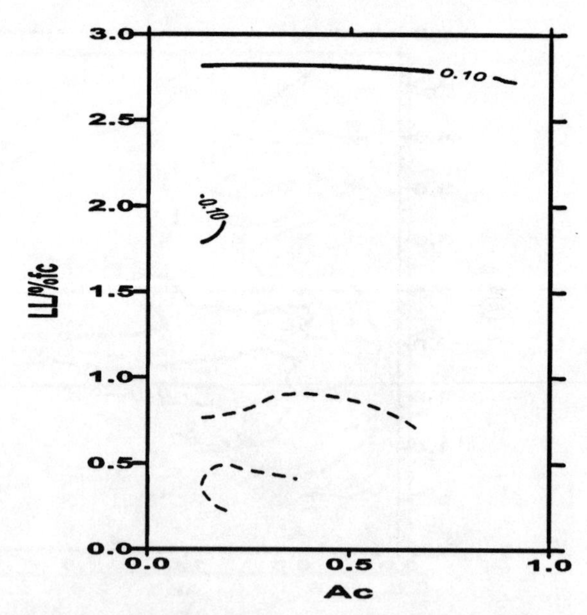

Figure 13. Predicted Soil Compression Index Values for Zone VI

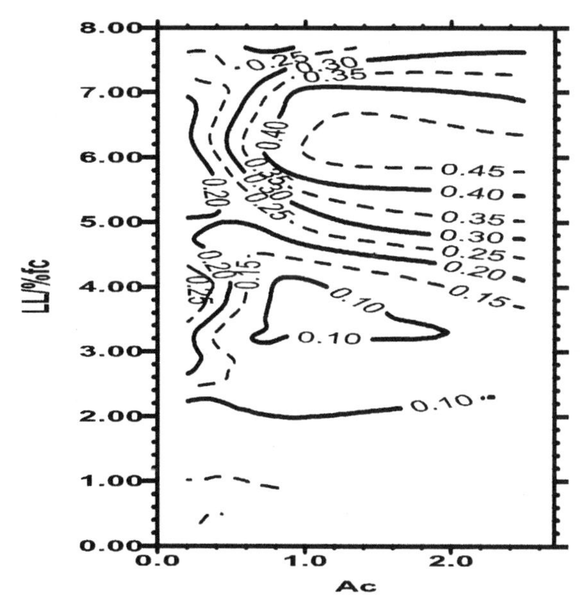

Figure 14. Predicted Soil Compression Index Values for Zone VII

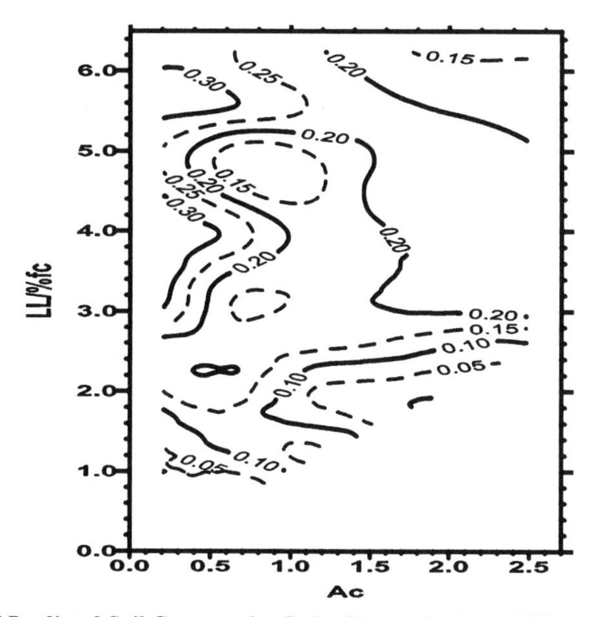

Figure 15 Predicted Soil Compression Index Values for Zone VIII

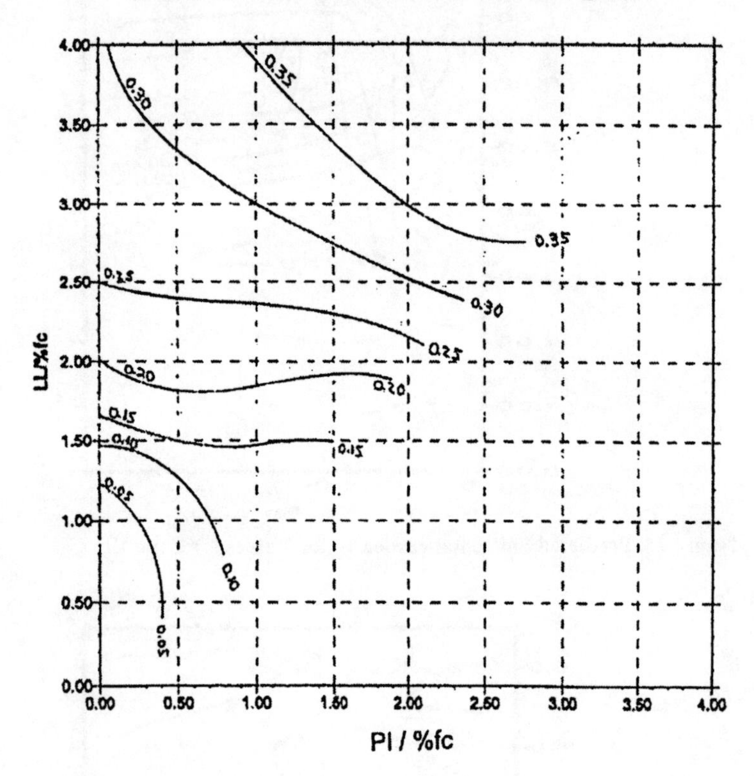

Figure 16a. Modified Soil Compression Index Surfaces from TBPE (2000)

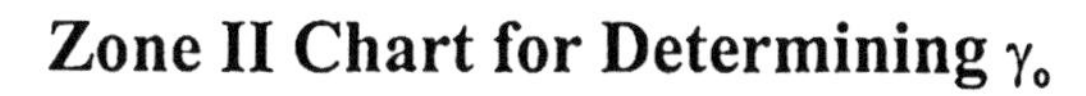

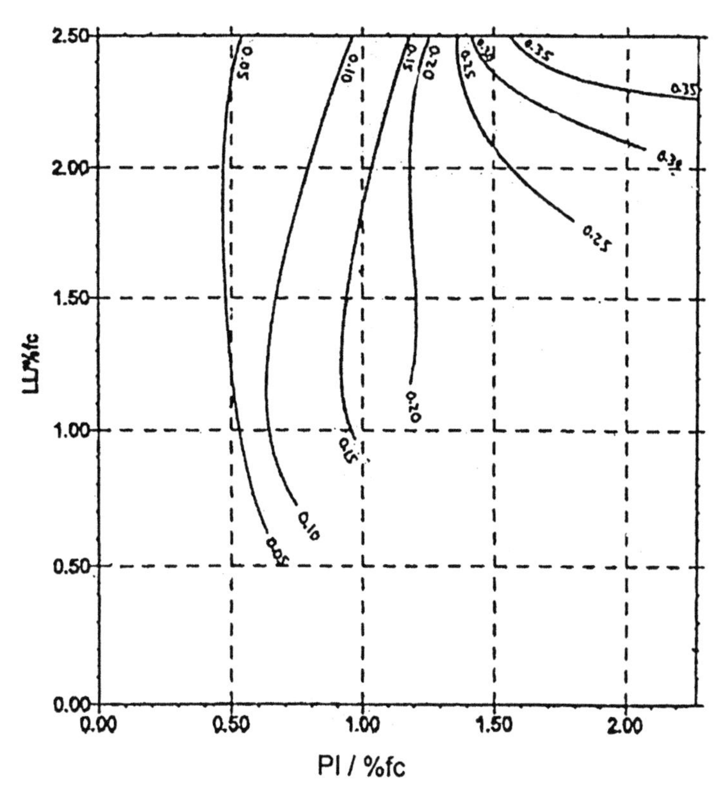

Figure 16b. Modified Soil Compression Index Surfaces from TBPE (2000)

Zone III Chart for Determining γ_o

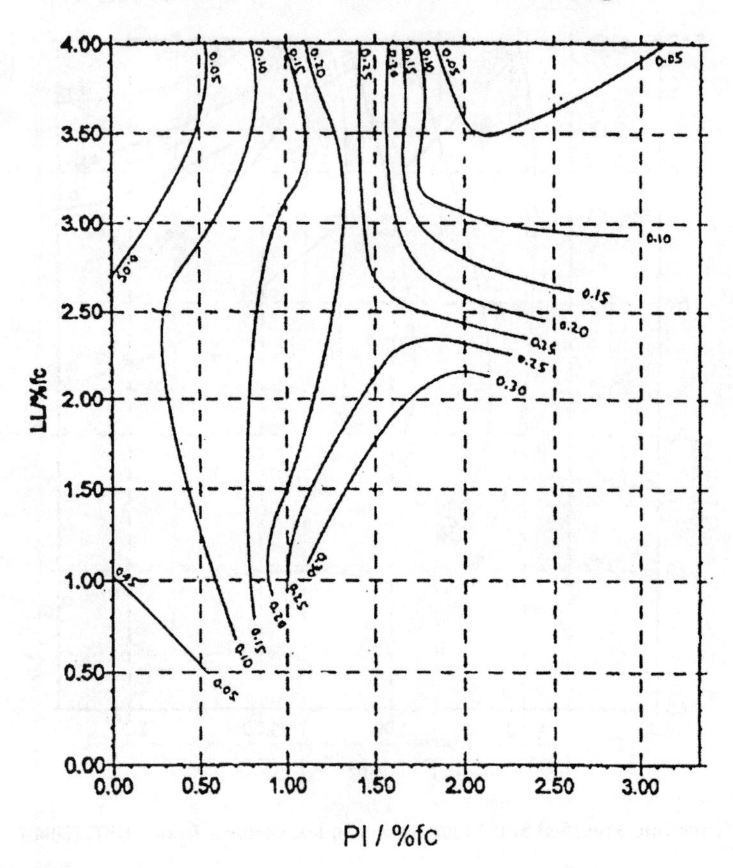

Figure 16c. Modified Soil Compression Index Surfaces from TBPE (2000)

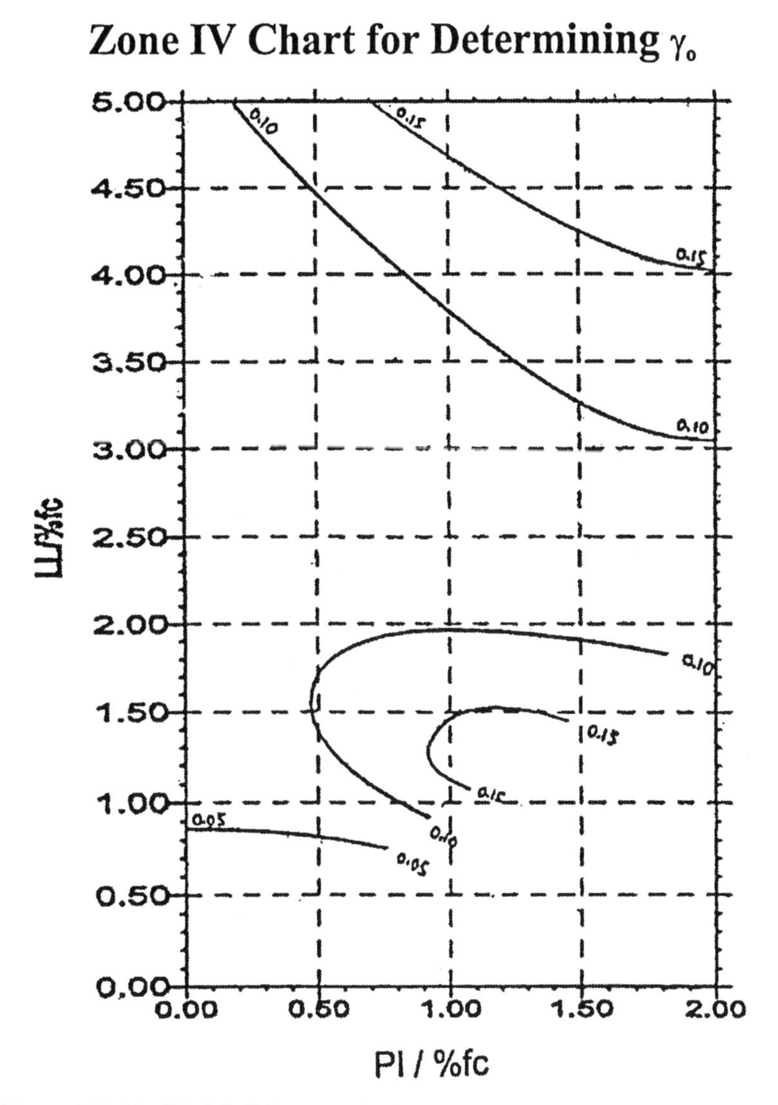

Figure 16d. Modified Soil Compression Index Surfaces from TBPE (2000)

SUPPORT PARAMETERS FOR SLABS ON GROUND ON EXPANSIVE CLAY WITH VEGETATION CONSIDERATIONS

By Kirby T. Meyer, Fellow ASCE* and Dean Read, Member ASCE**

Abstract

The background and procedures embodied in the recent Texas Board for Professional Engineers Advisory Report on Residential Foundations Appendix A represents extensive geotechnical research and practice relating to unsaturated clay soils. The effect of local soil properties and environmental factors on new design and forensic analysis should be a part of geotechnical, structural and foundation engineer's tool kits. These procedures, as well as the effects of barriers, drainage and vegetation on foundation support conditions are discussed and illustrated. Procedures for determining soil support parameters (E_m) and (Y_m) values and examples of designs will be presented.

Introduction

Foundations for single family, multi-family, and low-rise commercial buildings in many parts of the United States typically use shallow, stiffened, reinforced slab-on-ground concrete foundations for economy and speed of construction. The support parameters determined by the methods presented are equally applicable to all such foundations, whether reinforced with conventional re-bar steel or post-tensioning, which are designed for expansive soil sites. Factors influencing the support parameters for these foundations include the soil profile, climatic conditions, drainage, and vegetation.

Current procedures for obtaining the soil support conditions necessary for design, which are widely recognized, include that found in the Building Research,

----

*President and Senior Consultant, MLAW Consultants & Engineers, Austin, Texas, 512-835-7000, ktmeyer@mlaw-eng.com

**Vice-President and Senior Programmer, Geostructural Tool Kit, Inc., Austin, Texas, 512-835-7206, drread@gtksoft.com

the Wire Reinforcing Institute procedure (1991) and the Post Tensioning Institute manual "Design and Construction of Post-Tensioned Slabs on Ground" (PTI 1996). These three procedures have been or presently are recognized in national building codes and American Concrete Institute publications. Other local procedures are used, such as the "loss of support" method (Jacot 1990) in several western states, PVR (Determination of Potential Vertical Rise, 1995) evaluation and others. Only the first three have been nationally adopted and recognized in building codes. Of all procedures with which the authors are familiar, only the PTI procedure attempts to provide an analysis model that rationally accounts for the soil physics factors that influence the foundation support parameters. All other procedures are empirical in nature and give greatly different results for identical site, climatic, and soil profile data. Experience has shown that design results can be un-conservative.

The PTI procedure also has shortcomings, such as no specific guidance for drainage anomalies or vegetation effects and insufficient accounting for local soil properties and soil fabric.

The Texas Board for Professional Engineers Advisory Report (TBPE 2000) included Appendix A-Soil Support Parameters For Shallow Foundations On Expansive Soil Sites. The information contained in this appendix is an outgrowth of research by Lytton (1994), the senior author's experiences with design and observation of residential and commercial foundation slabs over a 30 year period, plus a vast new collection of soil testing data from the Natural Resources Conservation Services Laboratory, Lincoln, Nebraska (1999), which can be used as a replacement for the original charts and analysis used in the PTI Manual for deriving soil support parameters on expansive clay. The structural procedures in the PTI Manual can be utilized in producing designs using the proposed support parameters. The proposed new procedures for determining the PTI Support Parameters have not been adopted by the Post-Tensioning Institute.

The new procedures take into account environmental factors (Thornthwaite Index) as well as local conditions and soil properties for development of the commonly used four factors describing slab on ground support conditions as used in the PTI Manual procedures (Em-ctr, Em-edg, Ym-ctr, Ym-edg). The overall result is generally more conservative foundation designs, but not always. The new procedures were tested against the PTI Manual soil support procedures, the BRAB procedures and the WRI design procedures. The comparisons were based on soil properties of sixteen soil samples from around Texas and Louisiana, deriving soil support parameters and producing hypothetical designs for the same loads and slab rectangle. The results of this comparison may be seen in Meyer (2001). In the senior author's consulting practice, the principles embodied in the presented procedures have been used for the design of over 10,000 foundations.

The procedures presented represent a more complete methodology using unsaturated soil mechanics formulations for design of shallow foundations, based on extensive new test data, in a procedure that is simple and rapid to utilize. The use of the procedures fully benefits from a more careful gathering and consideration of geotechnical field and laboratory data, something that is critical to the design of these types of foundations, and long overdue.

Screening for Applicability

Sites for which this procedure is applicable should meet either of the following criteria:

- Soil of 0.6 m (2 ft) or more classified as CL or CH by the Unified Soil Classification System (1954) having a Plasticity Index of 20 or greater within the upper 1.52 m (5 ft) of the soil profile.
- PVR (TxDOT 124E 1995) calculated for the upper 2.7 m (9 ft), using "Dry" initial conditions, which is 2.5 cm (1.0 in) or greater.

If neither of these criteria is met, the foundation should be designed as a non-stiffened slab foundation, such as the BRAB or PTI Type II foundations, and the following procedures are not applicable.

Edge Moisture Variation Distance (E_m)

The edge moisture variation distance is the horizontal distance beneath the edge of a shallow foundation within which soil volume will change due to wetting or drying influences around the perimeter. In an edge lift case, the expansion in the soil is higher at the edges than in the center. The center lift case is one in which the soil volume is reduced at the edges relative to the center. A major factor in determining the moisture variation distance is the unsaturated diffusion coefficient, α. This, in turn, depends on the level of suction, the permeability, and the fabric of the soil. For the same diffusion coefficient and climatic factors, and assuming no vegetation influence, the E_m value will be larger for the center lift case in which moisture is withdrawn around the edges of the foundation. The E_m value will be smaller for the edge lift case in which moisture is drawn beneath the building into drier soil. Roots, layers, fractures or joints in the soil will increase the diffusion coefficient and the E_m value for both the edge lift and center lift conditions.

Using representative values based on laboratory test results in each significant layer, the following data are required to determine edge moisture variation distance, E_m, and differential movement, Y_m:

- Liquid Limit, LL
- Plastic Limit, PL
- Plasticity Index, PI
- Percentage of total sample soil passing No.200 sieve (% -No.200)

- Percentage of total sample soil finer than 2 microns (% -2 microns) expressed as a percentage of the total sample.
- Percent fine clay $\left(\frac{\% - 2microns}{\% - \#200}\right)$ reported as %fc, as percentage.

Calculate γ_h

$\gamma_h = \gamma_o$ (%fc / 100)

Terms: γ_o is the change of soil volume for a change in suction for 100% fine clay content. γ_h is the correction of γ_o for the actual % of fine clay (%fc).
Where, γ_o is determined using the following steps:

Step 1 Determine Mineral Classification Zone I, II, III or IV from Figure 1. If data does not fall within one of the four zones, use the nearest zone.
Step 2 Proceed to the chart corresponding to the zone determined in Step 1 to determine γ_o (Figures 2 through 5)
Note: The charts referenced in Steps 1 and 2 above are based on analysis of Natural Resources Conservation Service Soils properties laboratory test results database by Covar and Lytton (1999).

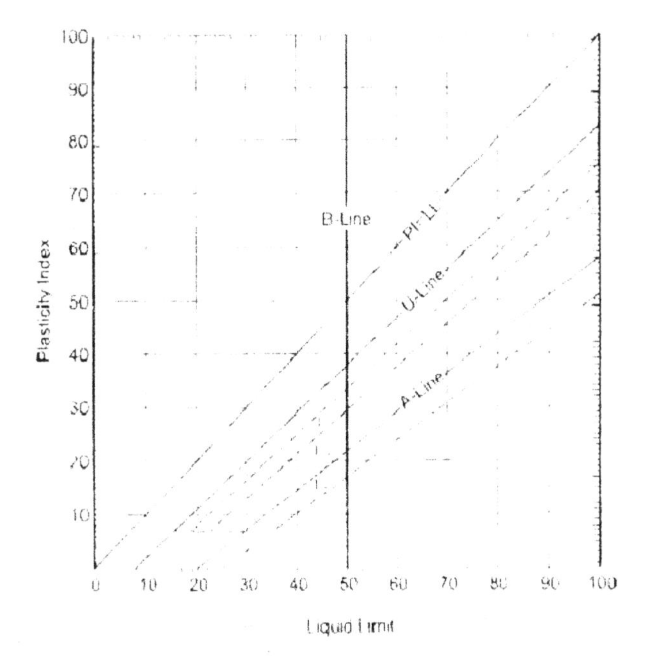

FIGURE 1. – MINERAL CLASSIFICATION CHART

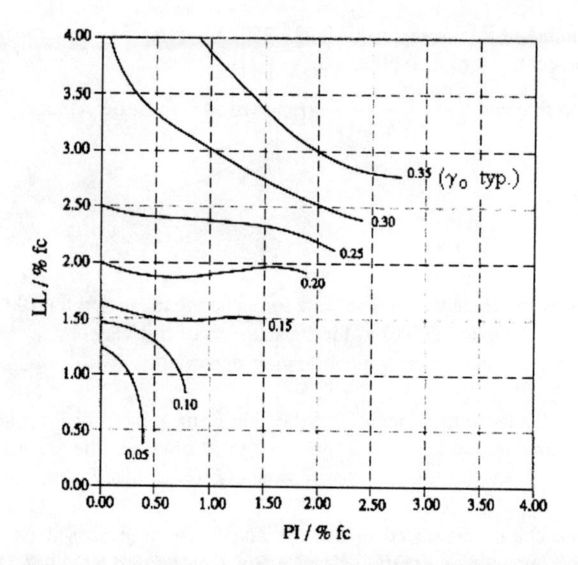

FIGURE 2. – ZONE I CHART FOR DETERMINING γ_o

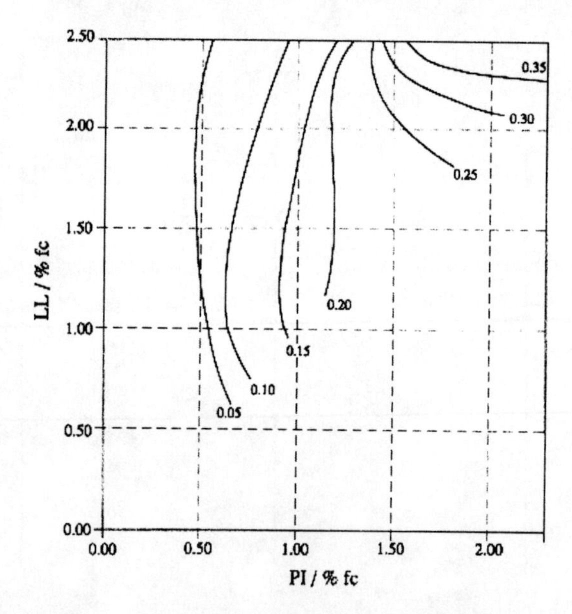

FIGURE 3. – ZONE I CHART FOR DETERMINING γ_o

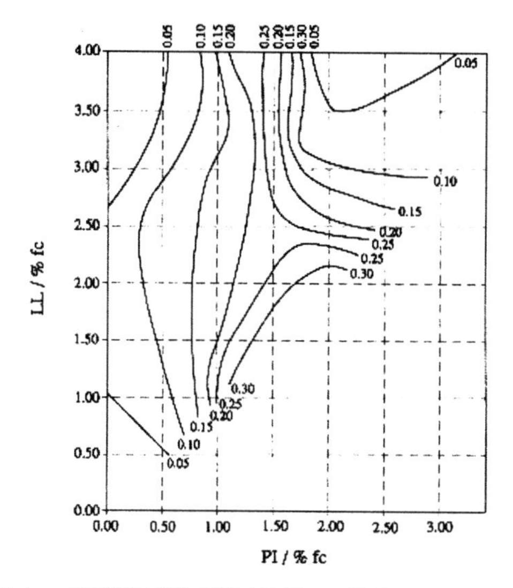

FIGURE 4. – ZONE III CHART FOR DETERMINING γ_o

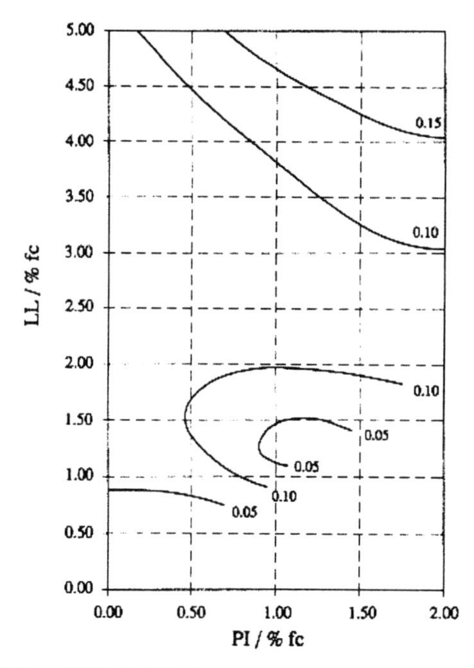

FIGURE 5. – ZONE IV CHART FOR DETERMINING γ_o

Calculate the unsaturated diffusion coefficient, α

$\alpha = 0.0029 - 0.000162\,(S) - 0.0122\,(\gamma_h)$

where $S = -20.29 + 0.1555\,(LL) - 0.117\,(PI) + 0.0684\,(\%\ \text{-No.200})$

The resulting unsaturated diffusion coefficient, α, for each significant layer, should be converted to the modified unsaturated diffusion coefficient, α', using Ff. $\alpha' = \alpha\,Ff$, where Ff is the soil fabric factor from Table 1 below:

TABLE 1. - SOIL FABRIC FACTOR

Condition	Ff
Soil profiles contain **few** roots, layers, fractures or joints no more than 1 per vertical 30 cm (1 ft)	1.0
Soil profiles contain **some** roots, layers, fractures or joints 2 to 4 per vertical 30 cm (1 ft)	1.3
Soil profiles contain **many** roots, layers, fractures or joints 5 or more per vertical 30 cm (1 ft)	1.4

The modified unsaturated diffusion coefficient, α', should be calculated for each significant soil layer to a minimum depth of 2.7 m (9 ft) by the procedure outlined above. Depths greater than this may be used if justified by geotechnical analysis. The evaluation of the edge moisture variation requires using a weighted average of the modified unsaturated diffusion coefficient with a weight of three for the top one-third, two for the next third and one for the bottom third of the selected profile of influence.

Determine edge moisture variation distance, E_m for both center lift and edge lift from Figure 6. (Multiply E_m by 0.304 to obtain values in meters.)

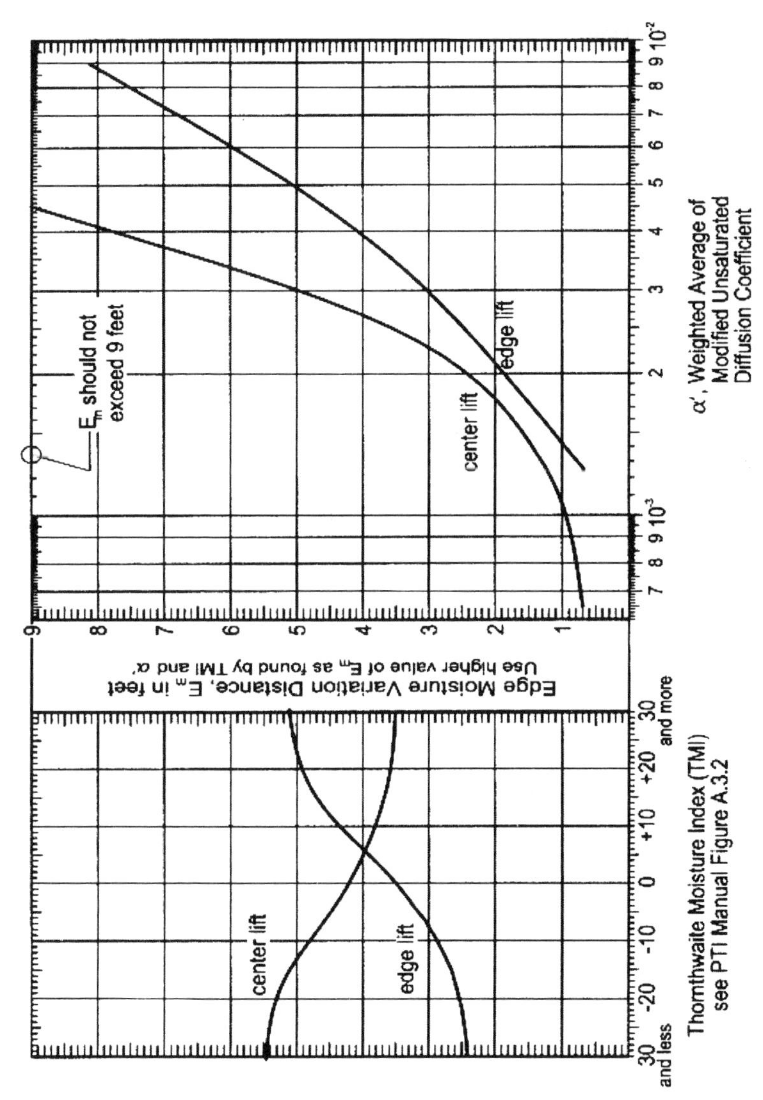

FIGURE 6. – DETERMINATION OF EDGE MOISTURE VARIATION DISTANCES

Differential Soil Movement (Y_m)

Differential soil movement (Ym) should be calculated using the change in soil surface elevation at two locations separated by a distance E_m within which the differential movement will occur. An initial and a final suction profile should be used at each of the two locations to determine differential movement. The final suction profile at each location should be determined from controlling suction conditions at the surface. A typical vertical suction profile is computed by using the principles of steady state unsaturated flow which links the controlling suction values at the soil surface to the controlling suction below the surface. The applicable principles of steady-state unsaturated flow may be found in Lytton (1994).

Procedures for estimating differential soil movements should be computer methods used to generate the design values of Y_m for the edge lift and center lift conditions. A simplified chart method is available in TBPE (2000) for non-layered soil profiles, but the effects of layered profiles, vegetation, barriers, drainage anomalies or plumbing leaks can be better handled using 2-d computer methods.

Vegetation, Barriers, Drainage and Plumbing Leaks

Effects of vegetation such as trees on the suction levels below a slab foundation can be severe. A tree removed from the slab area prior to construction can leave a zone of elevated suction that will gradually draw water into itself from surrounding soil, causing an area of unexpected heave after construction. Trees may also be planted or allowed to remain near the slab perimeter. Roots can penetrate beneath the slab, withdrawing water and causing a loss of support due to soil shrinkage.

Vertical moisture barriers may be used to reduce the soil support parameters (E_m and Y_m) provided the barriers are properly designed to virtually stop moisture migration to or from the under slab area on a permanent basis. Combined with proper details, they can also isolate the under slab soil from exterior tree root penetration.

The effects of a plumbing leak, barriers, and vegetation on distribution of soil suction may be seen in Figures 8, 9 and 10. The impact of vertical barriers on Em and Ym used in new design is also illustrated. The calculations for these charts were done using the computer program VOLFLO (1996). The results of the computer runs do not exactly match those that might be obtained using the simplified table and procedures found in TBPE (2000) due to slightly different formulations and assumptions. Comparisons of stiffening beam sizes and layout are presented for average environmental case, barrier cases, plumbing leak case and tree cases. See Tables 2 and 3 and Figure 7.

Cases 1 through 6 and 9 were analyzed using the VOLFLO program. Cases 7 and 8 were analyzed using a one-dimensional analysis for elevation change at a point. All calculations assumed climatic factors, soil properties, slab dimensions and slab loading to be the same. See Tables 4 and 5.

TABLE 2. – SUMMARY OF ANALYSIS CASES AND RESULTS

		FINAL	
CASE	DESCRIPTION	Em(m)(ft)	Ym(cm)(in)
1.	Typical environmental conditions, center lift, Ff 1.0	1.8 (5.8)	2.3 (0.89)
2.	Typical environmental conditions, center lift, Ff 1.4	2.7 (9.0)	2.3 (0.89)
3.	Typical environmental conditions, edge lift, Ff 1.0	1.0 (3.3)	7.8 (3.06)
4.	Typical environmental conditions, edge lift, Ff 1.4	1.4 (4.7)	7.8 (3.06)
5.	Vertical barrier to 0.6 m (2 ft.), center lift, Ff 1.4	2.7 (9.0)	0.8 (0.33)
	edge lift	1.4 (4.7)	4.2 (1.66)
6.	Vertical barrier to 1.5 m (5 ft) center lift, Ff 1.4	2.7 (9.0)	0 (0.00)
	edge lift	1.4(4.7)	0.7 (0.28)

CASE	DESCRIPTION	Ym Differential
7.	Plumbing leak below slab	9.9 cm (3.9 in)
8.	Pre-existing tree removed from under slab	9.4 cm (3.7 in)
9.	Tree remains outside slab, roots penetrating 1.8 m (6 ft) under slab	7.9 cm (3.1 in)

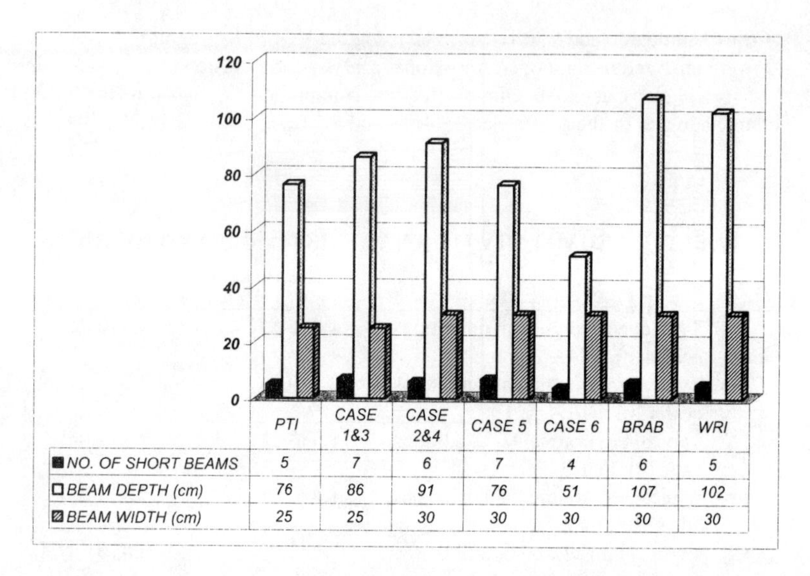

FIGURE 7. – COMPARISON OF DESIGNS USING VARIOUS METHODS

TABLE 3 – COMPARISON OF DESIGNS FOR VARIOUS METHODS

	No. of Short Beams	Beam Depth (cm)(in)	Beam Width (cm)(in)	Beam Reinforcing.
PTI	5	76(30)	25(10)	1 Tendon
CASE 1&3	7	86(34)	30(12)	1 Tendon
CASE 2&4	6	91(36)	30(12)	2 Tendons
CASE 5	7	76(30)	30(12)	1Tendon
CASE 6	4	51(20)	30(12)	1 Tendon
BRAB	6	107(42)	30(12)	4-#8 Top, 4-#9 Bottom
WRI	5	102(40)	30(12)	2-#6 Top, 2-#7 Bottom

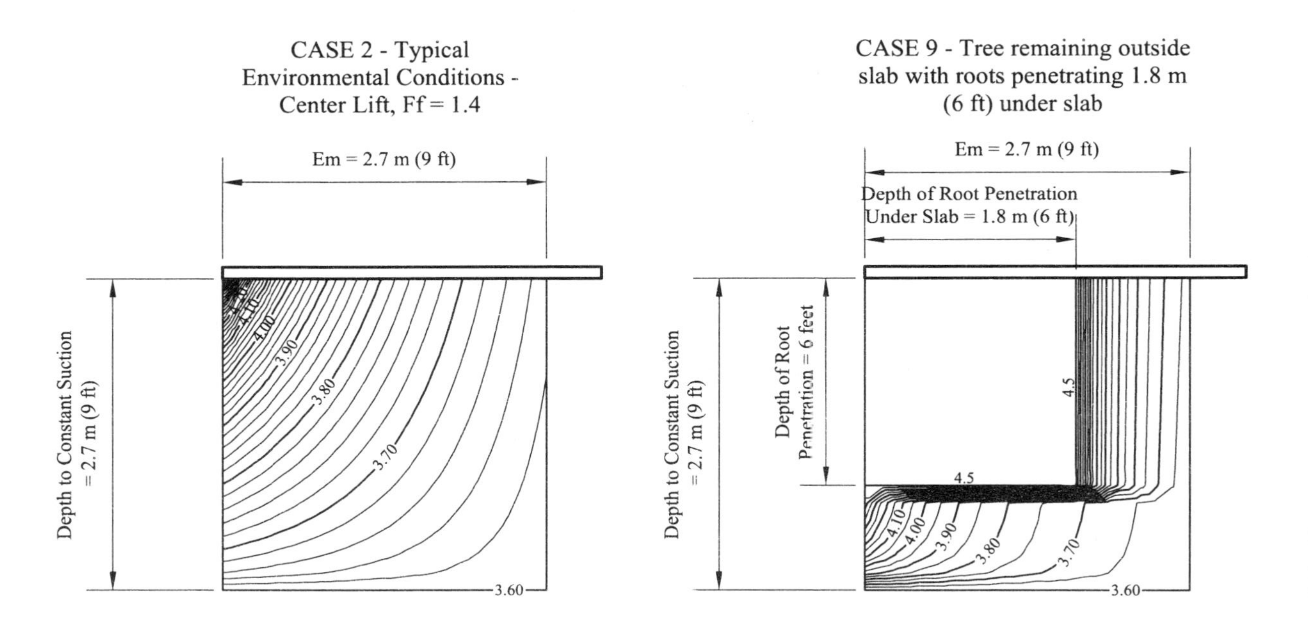

FIGURE 8. - CONTOURS OF EQUAL SUCTION FOR CASE 2 AND CASE 9

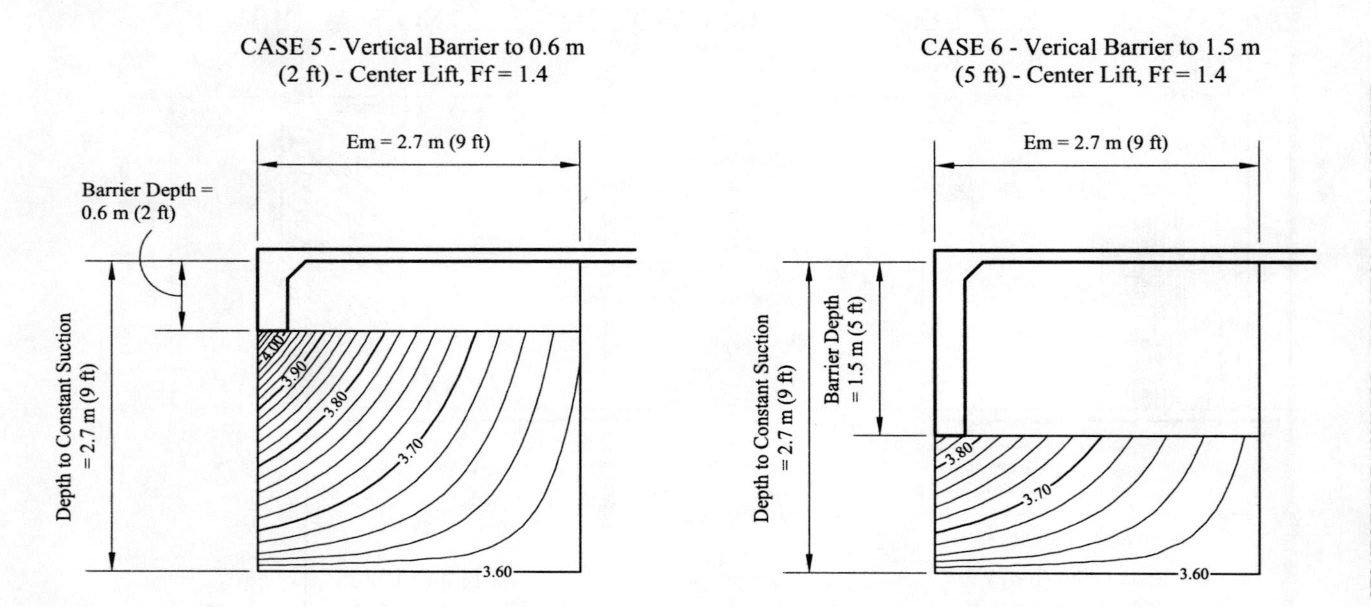

FIGURE 9. - CONTOURS OF EQUAL SUCTION FOR CASE 5 AND CASE 6

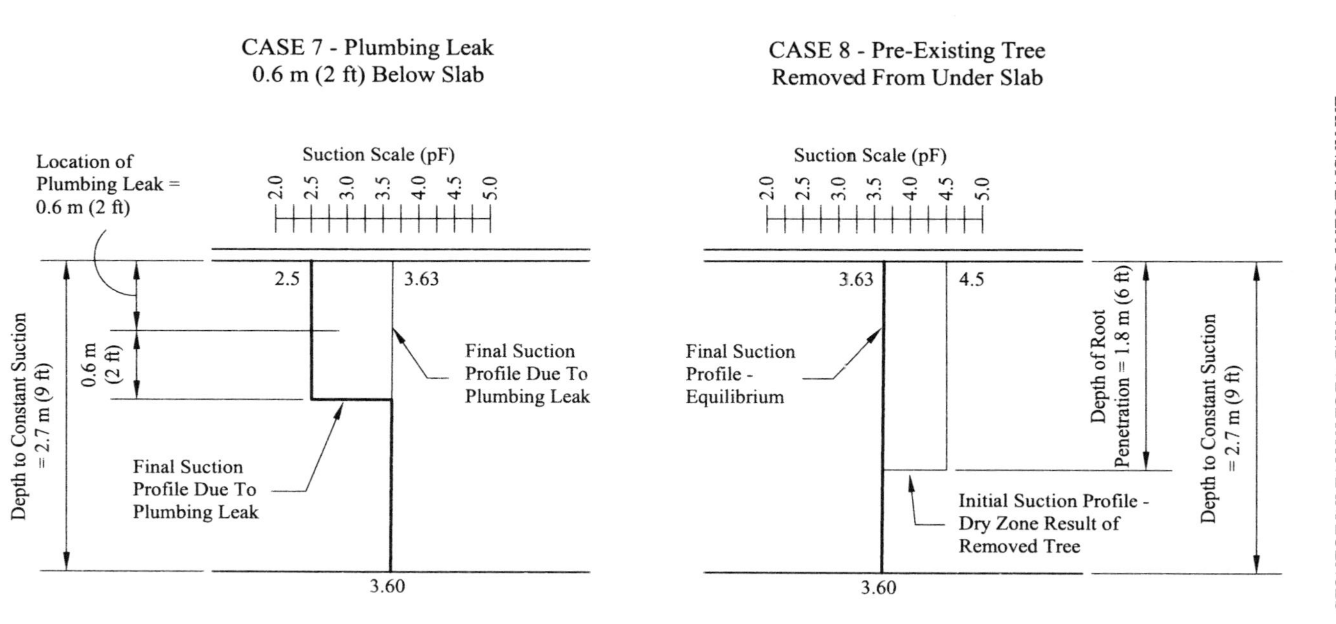

FIGURE 10. - SUCTION PROFILES FOR CASE 7 AND CASE 8

TABLE 4. – GENERAL ASSUMPTIONS

	Geometry	Length	18.3 m (60 ft)
		Width	13.7 m (45 ft)
		Slab Thickness	10 cm (4 in)
		Minimum Beam Depth	46 cm (18 in)
		Maximum Beam Depth	117 cm (46 in)
		Minimum Beam Width	25 cm (10 in)
		Maximum Beam Width	36 cm (14 in)
	Material Properties	Concrete Compressive Strength	20.7 MPa (3000 psi)
		Concrete Unit Weight	2403 kg/m^3 (150 pcf)
		Concrete Creep Modulus	10342 MPa (1500000 psi)
		Steel Yield Strength	414 MPa (60000 psi)
PTI Assumptions			
	Material Properties	Tendon Strength	1862 Mpa (270 ksi)
		Tendon Diameter	1.3 cm (1/2 in)
	Prestress	Minimum Effective Prestress	827 kPa (120 psi)
		Prestress Loss	103 MPa (15 ksi)
		Max Eccentricity	7.6 cm (3.0 in)
	Soil Properties	Sub Grade Friction Coefficient	0.75
		Soil Modulus of Elasticity	6.89 MPa (1000 psi)
	Load & Deflection	Deflection Criteria - Center	L / 480
		Deflection Criteria - Edge	L / 960
		Uniform Superimposed Load	195 kg/m^2 (40 psf)
		Perimeter Total Load	1339 kg/m (900 lb/ft)
BRAB Assumptions			
	Geometry	Beam Width	30 cm (12 in)
	Load & Deflection	Deflection Criteria	L / 360
		Dead Load	732 kg/m^2 (150 psf)
		Live Load	342 kg/m^2 (70 psf)
WRI Assumptions			
	Geometry	Beam Width	30 cm (12 in)
	Load & Deflection	Deflection Criteria	L / 360
		Weight of House	976 kg/m^2 (200 psf)

TABLE 5. – SOIL DATA USED IN ANALYSIS

Sample from Austin, Texas - Eagle Ford geologic formation

Liquid Limit	75
Plastic Limit	24
Plasticity Index	51
% passing #200	88%
% passing 2 μ	63%
% fine clay	72%

VOLFLO Assumptions

Depth to Constant Suction		2.7 m (9 ft)
Constant Suction Value		3.6 pF
Depth above which no Volume Change occurs		40 cm (1.31 ft)
Volume Change Guide Number		0.11
Velocity		
	Center Lift – Em=2.7 m (9.0 ft)	0.6 cm/mo (0.23 in/mo)
	Center Lift – Em=1.8 m (5.8 ft)	0.9 cm/mo (0.36 in/mo)
	Edge Lift – Em=1.4 m (4.7 ft)	-25.9 cm/mo (-10.2 in/mo)
	Edge Lift – Em= 1.0 m (3.3 ft)	-36.8 cm/mo (-14.5 in/mo)
Velocity Distribution Factor		0.5
Lateral Earth Pressure Coefficient		
	Drying	0.33
	Wetting	0.67
Vertical Volume Change Coefficient		
	Drying	0.5
	Wetting	0.8
Tree Root Penetration under foundation (Case 8 and 9)		1.8 m (6 ft)

CONCLUSIONS

Soil support parameters for slabs on ground foundations may be determined with greater accuracy than by currently available methods. The proposed methodology accounts for climatic factors as well as the local soil properties for obtaining the E_m and Y_m values used in the Post-Tensioning Institute Manual (PTI 1996) design procedures. Proper application requires more professional geotechnical investigations to account for important variables than has frequently been applied to these type foundation designs.

The proposed procedures also make possible the modeling of effects of barriers, vegetation and plumbing leaks on these types of foundations. The reduction of Y_m due to a vertical barrier is shown to be very significant.

The impact on design configurations (number and depth of stiffening beams and reinforcing) using the proposed procedures compared to the currently nationally published PTI (1996), BRAB (1968) and WRI (1991) methods can be seen in Figure 7 for the given hypothetical case. The new procedures designs are seen to be more or less conservative when compared to existing procedures, depending on soil properties or use of barriers.

Because of greater responsiveness to site properties, significant foundation cost savings can be realized by use of the proposed procedures, either in construction cost or avoidance of future claims for poor foundation performance.

REFERENCES

Building Research Advisory Board, Report #33. (1968). *Criteria for Selection and Design of Residential Slabs-on-Ground*, National Research Council, Washington, D.C.

Department of the Army, Corps of Engineer. (1954). "Unified Soils Classification", EM 1110-2-1803, Washington.

Jacot, Paul G. (1990). "A Practical Approach to Design and Construction of Post-Tensioned Slabs on Expansive Soils", locally published in California, Shaw & Co.

Lytton, R.L. (1994). "Prediction of Movement of Expansive Clays", ASCE Geotechnical Special Publication No. 40, Vol. 2, American Society of Civil Engineers, Reston, VA, 1827-1845.

Meyer, K.T. (2001). "Review of Procedures for Support Parameters for Stiffened Slabs on Ground on Expansive Clay", Spring 2001 Meeting, Proceedings, Texas Section American Society of Civil Engineers, Austin, Texas, 119-128.

National Soil Survey Center. (1999). Soil Laboratory Data Index, USDA, NRCS, Lincoln, NE. See following website for more information: http://www.statlab.iastate.edu/soils/nssh/631.htm#07. Statistical analyses by A.P. Covar and R.L. Lytton, personal communication.

Post-Tensioning Institute. (1996). *Design and Construction of Post-Tensioned Slabs-on-Ground*, 2nd Ed, Phoenix, AZ.

Texas Board for Professional Engineers. (2000). "Report of Design Sub-Committee of the Residential Foundation Committee", un-published local distribution, Austin, Texas. Copies are available from the author of this paper.

TxDOT Manual of Testing Procedures. (1995). Test Procedure 124-E, "Determination of Potential Vertical Rise", Austin, Texas.

VOLFLO. (1996). A computer program available through the Post-Tensioning Institute or Geostructural Tool Kit, Inc., Austin, Texas.

Wire Reinforcing Institute. (1991). "Design of Slabs-on-Ground Foundation", Leesburg, VA.

Investigating Field Behavior of Expansive Clay Soils

R. Gordon McKeen[1]

Abstract

Characterization based on samples obtained in conventional field investigations is an important step in obtaining engineering properties for use in selecting foundation type and making design decisions. A practical method for classification and estimating expansive soils behavior is described through application to three sites. The method is based on suction testing expansive clay samples and is suitable for routine use in geotechnical testing laboratories. In this paper, three cases are reviewed and the experience discussed. The use of relatively simple suction testing of soils is a valuable addition to routine geotechnical investigations of expansive soils.

Introduction

A classification system for use in evaluating the engineering behavior of expansive clay soils was published nine years ago, (McKeen, 1992). The procedures proposed at that time have been in use by the author in a variety of projects and found to be useful in practice. This paper presents some results from studies at three different sites to illustrate use of the techniques and the kind of information derived from this approach to the investigation of expansive soil conditions.

Classification System

Figure 1 illustrates the classification system based on suction testing of soil samples using the filter paper method, ASTM D 5298, (ASTM, 2000). Data shown are from Site 1 and represent suction and water content measurements on field samples taken at different depths in the soil profile. Boundaries for Categories I through V (dashed lines) were established based on experience in field investigations. Categories are shown in Table 1 and defined below. In the table the symbols are:

----

[1] Associate Director, ATR Institute, University of New Mexico, 1001 University Boulevard SE, Suite 103, Albuquerque, New Mexico 87106; phone 505-246-6001; gmckeen@unm.edu.

$\Delta h/\Delta w$ is the slope of the suction (h), water content (w) relationship, represented as the change in suction (Δh) divided by the change in water content (Δw)

C_h is the suction compression index, slope of volume change-suction relationship (McKeen, 1992; Lytton, 1994),

ΔH is the surface heave of a soil profile,

f is the lateral restraint factor, defined below,

s is a coefficient for heave reduction due to load.

Table 1. Proposed Expansive Soil Classification System

Category	$\Delta h/\Delta w$	C_h	ΔH* cm (in)	ΔH (%)	Remarks
I	> -6	-0.227	15.3 (6.0)	10.0	Special Case
II	-6 to	-0.227			High
	-10	-0.120	8.1 (3.2)	5.3	
III	-10 to	-0.120			Moderate
	-13	-0.040	2.7 (1.1)	1.8	
IV	-13 to	-0.040			Low
	-20	non-expansive	---	---	
V	< -20	non-expansive			Non-expansive

* ΔH for f = 0.5, active zone (Z_a) = 1.5 m (5 ft), Δh = 1.0 pF, s = 0.9

Category I soil is unusual and presents a serious problem for foundation design if present in significant amounts. If significant layers of Category I expansive soils are encountered, special testing is warranted to evaluate the site and develop recommendations for design and construction. In all known cases to date, these soils only occur as relatively thin bentonite seams. Category II soils are to most geotechnical engineers highly expansive clays. These require careful assessment of environmental, drainage and operational factors (possible sources of wetting or drying and loads) and development of designs to mitigate the effects of volume change. Category III soil is moderately expansive under normal conditions. Precautions in design and construction must be exercised to avoid difficulties, but severe heave/shrinkage are not typical except when exceptional moisture variation is imposed. Category IV soil does not exhibit behavior normally considered expansive. These soils do not require special considerations for shrink-swell behavior. Category V soil is non-expansive.

Table 1 shows the slope of the suction water content curve for each soil classification. The next column is the corresponding value of suction compression index derived from the models discussed below. A heave prediction (ΔH) in length units is the next column. This is computed based on the assumptions stated at the bottom of Table 1. The heave is then expressed as a percent of the soil profile involved. The last column is a name for the categories.

Figure 2 shows the same data presented in Figure 1 plotted as the slope of the suction-water content relationship computed for each data point versus the sample

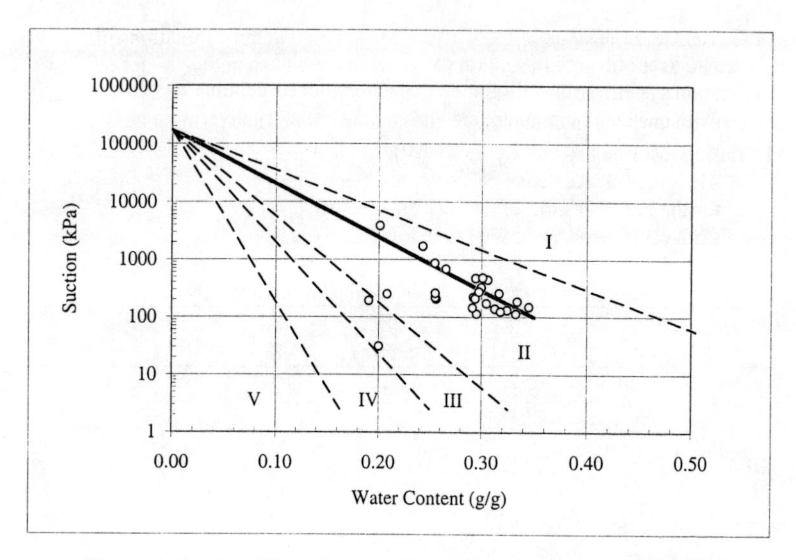

Figure 1. Suction-Water Content Classification System, Site 1 Data.

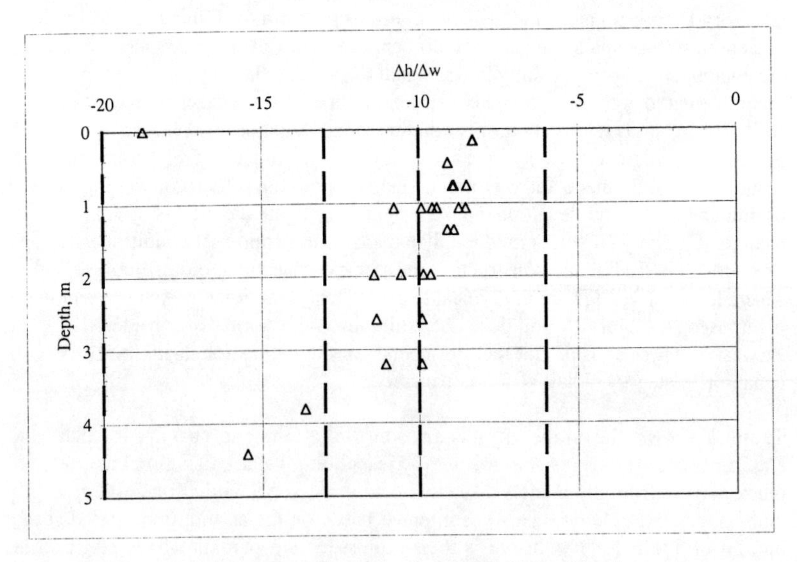

Figure 2. Soil Classification-Depth Data, Site 1.

depth below the surface. The slope is computed based on the assumption of a common intercept of 174390 kPa (6.25 pF) as illustrated in Figure 1. Based on the classification system, soils with flatter slope are more susceptible to volume changes with changes in moisture condition. The category dividing lines are plotted as vertical dashed lines in this case. For Site 1, the soil profile is composed of three distinct clay layers that can be identified in Figure 2. The top layer is highly expansive and extends to a depth of about 1.5 m. The second layer is moderately active and extends from 1.5 to 3.5 m. Below 3.5 m is a third layer with low expansion potential. Figure 3 shows the suction measured plotted versus depth for use in estimating the active zone depth, discussed below.

The data shown are from samples taken in three different borings at the same site. Clays were identified in boring logs consistent with the conclusion derived based on Figure 2. Soils were identified as dark gray clay, tan and gray clay, tan and gray sandy clay. Conventional data such as Atterberg Limits, grain size distribution and Unified Soil Classification also distinguish these soils. However, the above classifications system may be used to directly estimate surface heave/shrinkage of the soil at this site.

Heave Prediction

In the previous paper, McKeen (1992), an estimate of heave was developed using the following equation:

$$\Delta H = C_h \, \Delta h \, \Delta t \, f \, s \qquad (1)$$

where:

ΔH is the vertical movement at the surface,
C_h is the suction compression index for the layer (slope of the volume change versus suction relationship),
Δh is the suction change in the layer in pF,
pF is the exponent of negative water pressure in cm, Schofield (1935)
Δt is the thickness of the layer,
f is the lateral restraint factor, used to convert the volume change estimate to a vertical heave/shrinkage estimate:

$$f = \frac{[1 + 2K_o]}{3} \qquad (2)$$

@ $K_o = 1, f = 1$ @ $K_o = 0, f = 0.333$
[based on field data] $0.50 < f < 0.83$]
K_o is the coefficient of earth pressure at rest
s is a coefficient for load effect on heave
$s = 1.0 - 0.01(\% \text{ SP})$ [For % SP ≤ 50}
% SP is percent of swell pressure applied
s = 0.9 is assumed unless SP is measured.

The value of C_h may be obtained on the basis of testing. However, in routine geotechnical work this is cumbersome and time consuming. It may be estimated based on the correlation shown in Figure 4. The linear fit is based on experience and was originally published in 1992, it is represented by the following equation:

$$C_h = (-0.02763)*(\Delta h/\Delta w) - 0.38704 \quad (3)$$

The curve shown was developed based on more extensive testing over a wider range of materials, Perko, et. al (2000). It is represented by the following equation:

$$C_h = -10*(\Delta h/\Delta w)^{-2} \quad (4)$$

A value of C_h may be estimated by use of data in Figure 2 to find $\Delta h/\Delta w$. For example from surface to 1.5 m, -9, 1.5 to 3.5 m, -11 and below 3.5 m –14 are the respective values for $\Delta h/\Delta w$. On the basis of these values C_h may be estimated for each layer, -0.12, -0.08, and –0.05 respectively. The other variables in the Equation (1) must be determined for the site. Traditional geotechnical engineering has few tools for making these estimates.

The layer thickness is straight-forward and may be estimated from borings. The suction change and active zone depth for design are formidable geotechnical problems. The data is Figure 3 is helpful, but without understanding the climate, drainage and site use preceding the sampling, this can be misleading. An understanding of the site climate, drainage and use is one of the most difficult tasks in foundation design on expansive soils. In most situations a range of 98 kPa (3 pF) to 9800 kPa (5 pF) is reasonable for wet and dry boundaries for suction at the soil surface. The longer a drying event is prolonged the deeper the effects will reach, McKeen and Johnson (1990). A plot of suction versus depth is the most direct means of estimating the active zone depth, however, numerous factors must be assessed to develop a design estimate. Figure 3 presents suction versus depth data for Site 1. Since these data were obtained in the summer, drying at the surface is apparent. Depth of suction variation appears to be about 2 m for the conditions preceding the time of sampling. Normal equilibration will result in surface soils equilibrating with the soils at depth, in this case an equilibrium suction of about 250 kPa (3.5 pF) is indicated. The difficult question is whether during the design life these values (active zone 2 m, equilibrium suction 250 kPa) are appropriate. Table 2 below illustrates the importance of these in estimating heave behavior of the soils, using three different sets of values that may be reasonable under different circumstances. As shown, the results range from about 5 to 15 cm for this site.

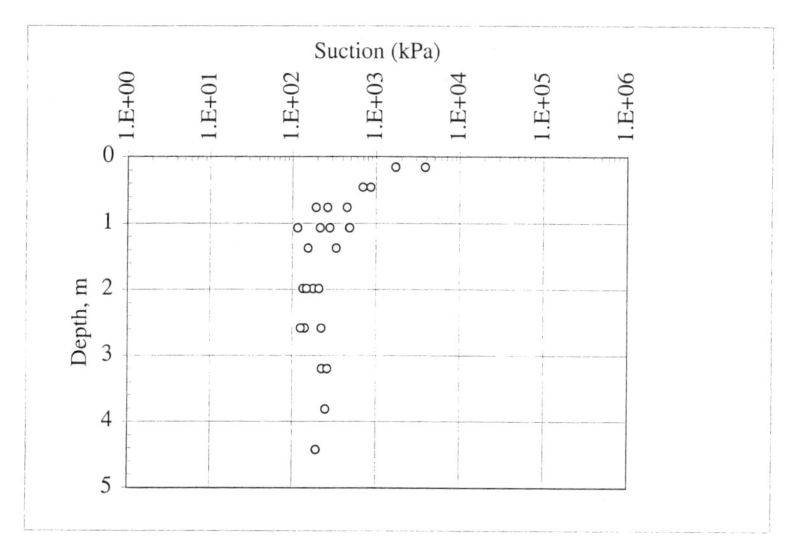

Figure 3. Suction Profile Data, Site 1.

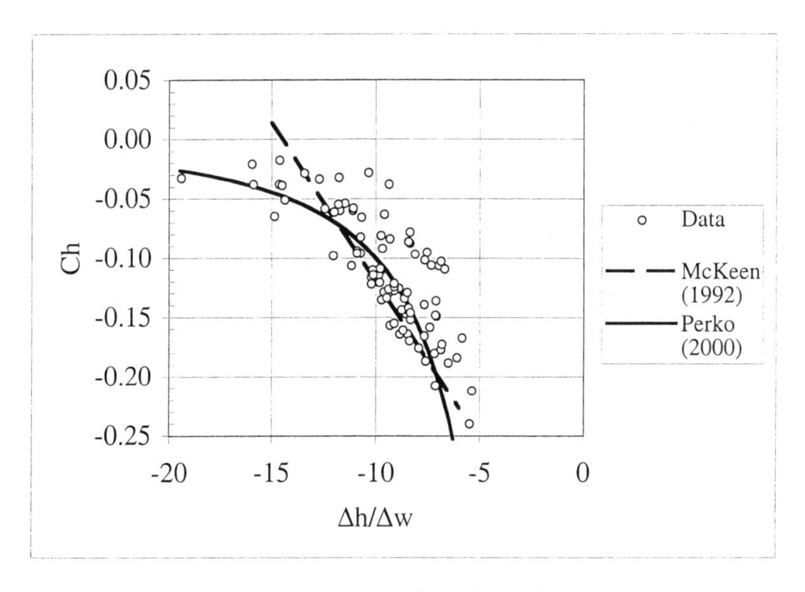

Figure 4. Estimates for Ch from h/w.

Table 2. Heave Estimates for Site 1.

Example	Active Zone m	Δt m	C_h	Δh log(kPa)	f	s	ΔH cm
I	0 to 1.5	1.5	- 0.12	0.5	0.5	0.9	4.1
	1.5 to 2.0	0.5	- 0.08	0.2	0.8	0.9	0.7
	Total						4.8
II	0 to 1.5	1.5	- 0.12	1.0	0.5	0.9	8.1
	1.5 to 2.0	0.5	- 0.08	0.2	0.8	0.9	0.7
	Total						8.8
III	0 to 1.5	1.5	- 0.12	1.0	0.5	0.9	8.1
	1.5 to 3.5	2.0	- 0.08	0.5	0.8	0.9	6.5
	Total						14.6

Site 1 Expansive Clay Marine Deposit.

Site 1 illustrates the estimated result of constructing a cover on the soil represented by the data in Figures 1, 2 and 3 isolating it from the climate and allowing it to equilibrate with soil below that affected by seasonal influences. Example II illustrates a greater suction change ($\Delta h = 1.0$), that may represent differing site moisture conditions at the time of construction (more drying) or some other changed condition between the time of sampling and the time of construction. Example 3 assumes a greater active zone as well as the greater suction change that may result from several different factors. The conclusion is there are three critical issues in making volume change predictions for design. They are (1) soil potential for volume change, indicated by C_h, (2) moisture condition change, indicated by suction change (Δh), and (3) amount of soil involved, indicated by the active zone or depth of wetting. Each of these characteristics must be carefully evaluated. It is imperative to realize the moisture conditions vary between sampling and construction and with this variation the depth of soil affected may also change significantly. Over the life of the facility the depth of soil subjected to moisture variation may vary considerably.

The investigation at Site 1 was made to evaluate performance of a general aviation airport pavement. The pavement exhibited significant heaving which produced roughness unacceptable for light aircraft operations.

Site 2 Deep Cut in Claystone.

Site 2 was investigated because of damages to shallow foundations supporting one and two story residential structures. Distress was variable from one location to the next and maximum soil heave was in the range of 20 to 25 cm. Site soils are claystone and it was apparent from examination of samples that sand seams occurred at about 1, 2 and 4 m. The bedding was at a steep angle so the sand seams were providing access for surface water to penetrate into the claystone. In Figures 5 and 6 data show the presence of very wet (low suction) layers. Note the claystone above and below the wet seam also exhibits reduced suction. On the basis of these data the soils at 4 m depth are undergoing moisture variation. Surface soils do not exhibit a drying or wetting trend because the boring was made through a paved area that had reached equilibrium. It appears based on review of the data the soils are being wetted by surface water through infiltration along the sand seams to depths of at least 4 m. The definition of active zone is complex because of uncertainty about the depth of wetting. The data are too shallow to be used to determine the value of constant suction for the site. Some assumed conditions are presented in Table 3 to illustrate the range of possible heave for this site. Since this is a complex site additional work is needed for design of shallow footings. On the basis of these data, it is clear that foundations will have to be very stiff if they are shallow or very deep to penetrate beyond the claystone layers affected by infiltration of surface water through the sand seams.

Table 3. Heave Estimates for Site 2.

Example	Active Zone (m)	Δt (m)	C_h	Δh log(kPa)	f	s	ΔH (cm)
I	0 to 1.0	1.0	- 0.04	0.5	0.5	0.9	0.9
	1.0 to 3.5	2.5	- 0.07	0.5	0.8	0.9	6.3
	Total						7.2
II	0 to 1.0	1.0	- 0.04	0.5	0.5	0.9	0.9
	1.0 to 3.5	2.5	- 0.07	0.5	0.8	0.9	6.3
	4.0 to 6.0	2.0	- 0.08	0.5	0.8	0.8	5.1
	Total						12.3
II	0 to 1.0	1.0	- 0.04	0.7	0.5	0.9	1.3
	1.0 to 3.5	2.5	- 0.08	0.7	0.8	0.9	10.1
	4.0 to 6.0	2.0	- 0.09	0.7	0.8	0.8	8.1
	Total						19.4

Site 3. Deep Fill in Expansive Clay.

Site 3 was investigated because of damage to shallow foundations to lightly loaded structures. Figure 8 shows suction-water content classification data. The soil varies from highly expansive to low expansion potential and the low expansion materials are wet. Based on the classification data in Figure 9 the soil is composed of layers that are distinct and alternate between low expansion potential (surface, 2 to 3 m, and 4 to 5 m) and highly expansive materials (1 to 2 m, 6.5 to 8.5 m). The low expansion materials are wet and based on suction profile data in Figure 10 water is infiltrating layers below. This site is a deep fill that was obviously constructed using a wide range of materials. It is difficult to judge the equilibrium condition for this profile because all of the material in the boring was fill. At the

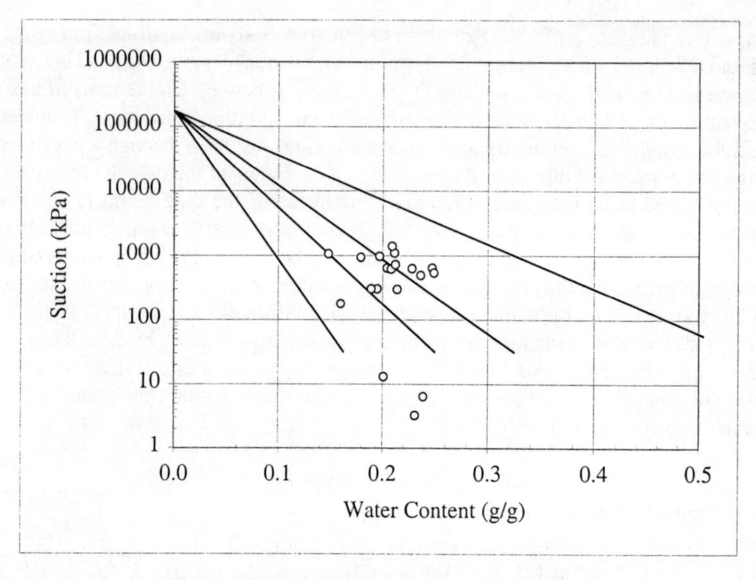

Figure 5. Suction Water Content Data, Site 2.

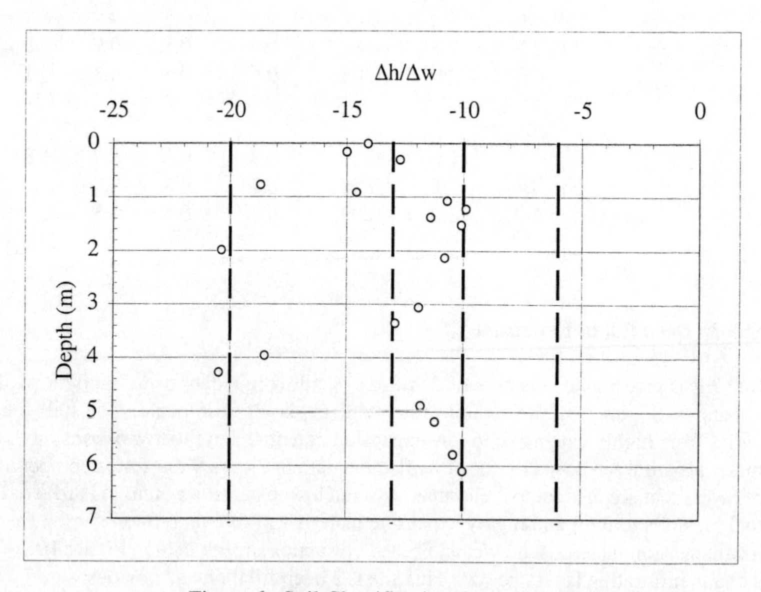

Figure 6. Soil Classification Data, Site 2.

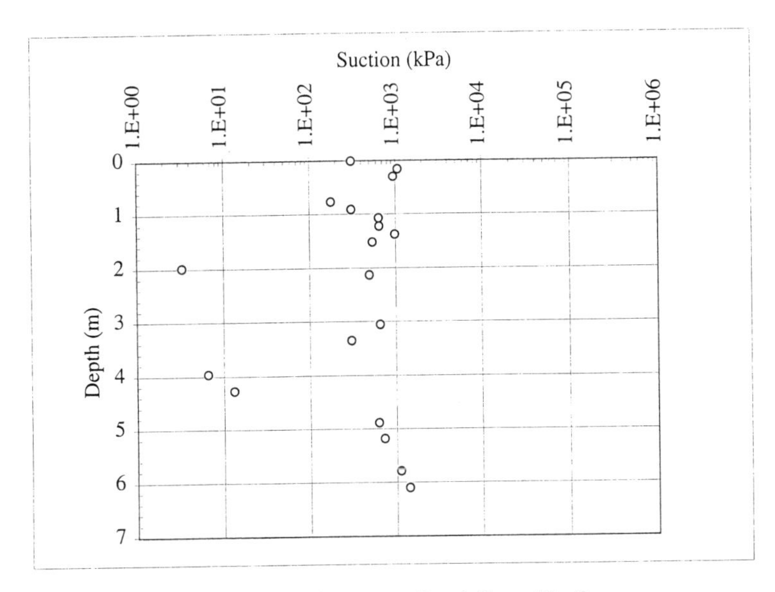

Figure 7. Suction versus Depth Data, Site 2.

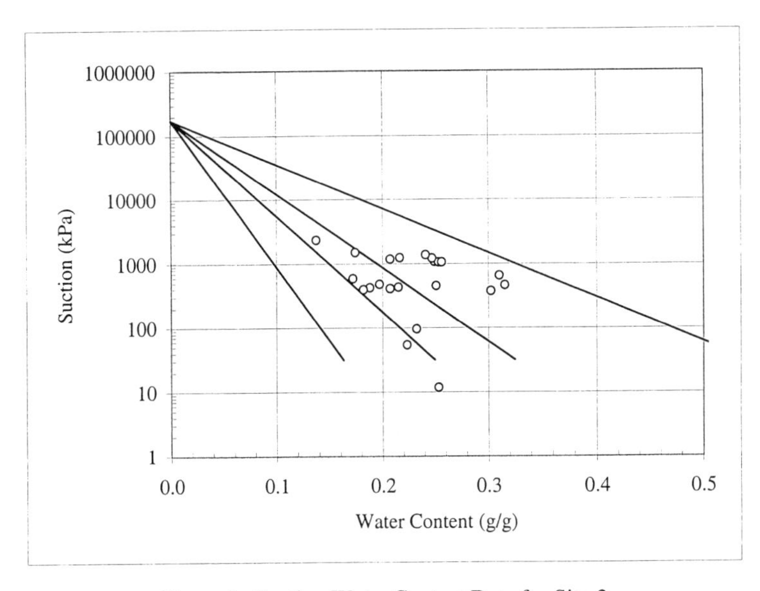

Figure 8. Suction Water Content Data for Site 3.

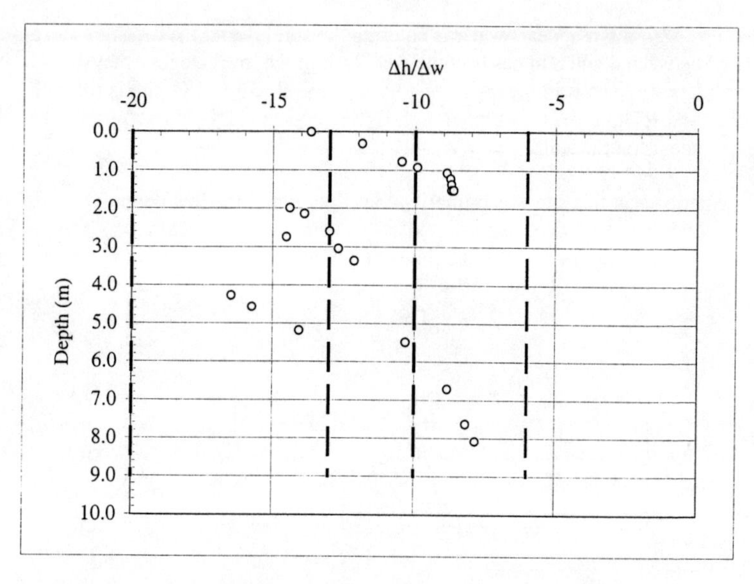

Figure 9. Soil Classification versus Depth, Site 3.

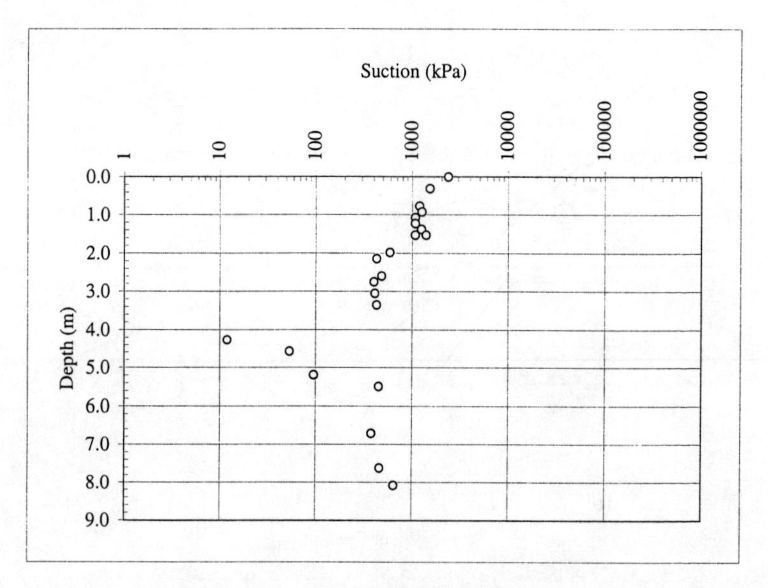

Figure 10. Suction Profile Data, Site 3.

time of this boring it appears water is entering the soil layer at the depth of 4 m and the soil between 4 and 8 m has become wet. Soil in the low category may consolidate while soils in the high category may swell. Since this site is fill, rather than natural soils, it is imperative to investigate density and obtain volume change measurements on the soils.

The experience at this site was primarily consolidation of the low potential expansion layer that had become wetted at depths of 3 to 5.5 m below the surface. Some near surface soils had exhibited heaving in areas where water was not controlled.

Summary and Conclusions.

Suction based testing offers the practicing geotechnical engineer a powerful tool for investigation of expansive soil behavior. Standard test methods are available and have been used in the context of practical investigations. By characterizing the soils using suction tests and plotting data as illustrated, several insights into soil behavior may be derived. Differences between soil layers are clearly described. The potential for moisture changes may be quantified. The potential for expansive behavior may be estimated which makes it possible to decide when more extensive soil behavior investigations are warranted. As these methods are introduced into the routine work of geotechnical engineers additional insight and experience will enhance their value.

The interaction of expansive soil and the surrounding environment is understood to some extent. The forces driving soil behavior are largely controlled by water. Suction is the best means of measuring the influence of water on soils. Further study is needed to establish guidelines for environmental factors and the effects of irrigation and leaking utilities. These effects can be quantified if careful investigations and proper measurements are made routinely and data base files created for use in guiding future designs and predictions of soil behavior.

References.

ASTM (2000), Test Method for Measurement of Soil Potential (Suction) Using Filter Paper, D 5298, Section 4 Construction, Volume 04.08, Annual Book of ASTM Standards, American Society for Testing and Materials.

Lytton, R. L., (1994) Prediction of Movement in Expansive Clays, *ASCE Geotechnical Special Publication No. 40, Volume 2, Vertical and Horizontal Deformations of Foundations and Embankments*, p.1827-1844.

McKeen, R. G. and Johnson, L. D. (1990), Climate-Controlled Soil Design Parameters for Mat Foundations, *Journal of Geotechnical Engineering* , ASCE, Vol. 116, No. 7, July, page 1073-1094.

McKeen, R.G., (1992), A Model for Predicting Expansive Soil Behavior, *Proceedings of the 7th International Conference on Expansive Clay Soils*, ASCE, Dallas, Texas, p-6.

Perko, H.A., Nelson, J.D. and Thompson, R.W. (2000), Suction Compression Index Based on CLOD Test Results, *in Advances in Unsaturated Geotechnics*, Geotechnical Special Publication Number 99, GeoDenver, ASCE, pp. 393-408.

Schofield, R. K. (1935), "The pF of the Water in Soil," *Transactions*, 3rd International Congress of Soil Science, Vol. 2, pp. 37-48.

DEPTH OF WETTING AND THE ACTIVE ZONE

John D. Nelson, P.E., Fellow, ASCE[1]
Daniel D. Overton, P.E., Member, ASCE[2]
Dean B. Durkee, P.E., Member, ASCE[3]

ABSTRACT

The term "active zone" generally refers to the zone of soil that either is contributing to or has the potential to produce heave. In order to predict heave at any particular time, it is necessary to define the zone of soil that has experienced an increase in water content and what the swell potential of that zone is. Different terms can be used for various zones that have in the past been used to define the "active zone". This paper discusses the usage of the term "active zone" and presents a definition that relates it to the movement of water in the soil profile.

The nature of water movement in soil, the definition of important zones that influence heave, and the interaction of different foundation elements with the expansive soil are shown to be interrelated. The depth of the wetting front in the soil profile is an important factor in prediction of heave and pier movement. In order to compute the maximum heave that can occur, it is necessary to assume that the depth of wetting will eventually extend throughout the depth of potential heave. Therefore, this depth represents the maximum depth of the active zone.

1. Professor, Civil Engineering Department, Colorado State University, Fort Collins, Colorado 80523, and Corporate Consultant, Shepherd Miller, Inc., 3801 Automation Way, Suite 100, Fort Collins, Colorado 80525, phone: 970-223-9600.
2. Senior Geotechnical Engineer, Shepherd Miller, Inc., 3801 Automation Way, Suite 100, Fort Collins, Colorado 80525, phone: 970-223-9600.
3. Senior Geotechnical Engineer, Golder Associates Inc., 2700 W. Central Avenue, Suite 300, Phoenix, Arizona, phone: 602-728-0400.

INTRODUCTION

Over the past two or three decades the term "active zone" has taken on several different meanings. These generally refer in some way to the zone of soil that either is contributing to or has the potential to produce heave of the ground surface. The two ingredients that cause heave are expansive soil and an increase in water content. Thus, the active zone must be related to that zone in which the water contents have changed or have the potential to change. As water migrates through soil, different zones of the soil profile become wetted at different times, some of which may have more swell potential than others. Consequently, the zone of soil that is contributing to heave varies with time.

In order to predict the amount of heave that will occur at a particular time, it is necessary to know what zone of soil is being wetted at that time, and the expansive nature of that soil. The effect of the heave on a structure founded on the soil will depend on the particular member of the foundation that is being affected by the heave, and where in the soil profile that member is located. In the following paper it will be shown that whereas movement of a slab-on-grade will begin very soon after heave commences, movement of piers founded in the same profiles may not begin until several years after slab movement begins.

Conservative design of a structure on expansive soil must consider the maximum amount of heave that can occur in the lifetime of the structure. Prediction of the maximum heave requires that the largest zone of expansive soil that can be wetted must be defined, variations of expansive soil properties in that zone must be determined, and the greatest heave potential of that zone must be predicted.

This paper will discuss the nature of water movement through soil, the definition of the important zones of soil profiles that influence heave, and the interaction of different foundation elements with the expansive soil. It will draw heavily on the experience of the authors relating to cases in the Front Range area of Colorado. The "Front Range area" as referred to herein is the area to the east of the foothills of the Rocky Mountains, and includes Fort Collins, Denver, Colorado Springs, and all points in between.

ACTIVE ZONE

Engineers have attempted to determine "active zone" depth using different definitions, each of which considers a particular emphasis. For purposes of consistency, the following four definitions will be used in this paper. The authors believe that these definitions are rigorous, and should be used universally. (Nelson, et al. 1994; Nelson, Durkee & Bonner, 1998; Durkee, 2000)

1. **Active Zone** (Z_a) is that zone of soil that is contributing to heave due to soil expansion at any particular time. The active zone will normally vary with time.

2. **Zone of Seasonal Moisture Fluctuation** (Z_s) is that zone of soil in which water contents change due to climatic changes at the ground surface.

3. **Depth of Wetting** (Z_w) is the depth to which water contents have increased due to the introduction of water from external sources, or due to capillarity after the elimination of evapo-transpiration. The external sources can include such things as irrigation, seepage from ponds or ditches, broken water lines, and others.

4. **Depth of Potential Heave** (Z_p)is the depth to which the overburden vertical stress equals or exceeds the swelling pressure of the soil. This represents the maximum depth of Active Zone that could occur.

MOVEMENT OF WATER IN SOILS

Expansive soil problems occur primarily in arid or semi-arid climates. An idealized water content profile in a uniform soil at an undeveloped site in a dry climate is shown by profile A in Fig. 1. (Nelson, et al. 1994; Durkee, 2000) Below some depth (Depth Z_s) an equilibrium water content exists. Above Depth Z_s, the water content decreases due to water losses from the ground surface, usually evapo-transpiration. If a cover is placed on the ground surface that is large enough that edge effects can be neglected, surface water losses are eliminated, and the water content profile will come into equilibrium with the environment as shown, for example, by Profile B. If the ground surface is subjected to temperature fluctuations such as due to summer and winter climates, the water contents in the zone affected by temperature changes will fluctuate about Profile B. Profile C, for example, would be typical for winter conditions and Profile D would be typical for summer conditions. The zone in which temperature effects occur, and depths below that in which climatic effects can change the water content define the zone of seasonal fluctuation. The depth of this zone would be less than or equal to the depth Z_s.

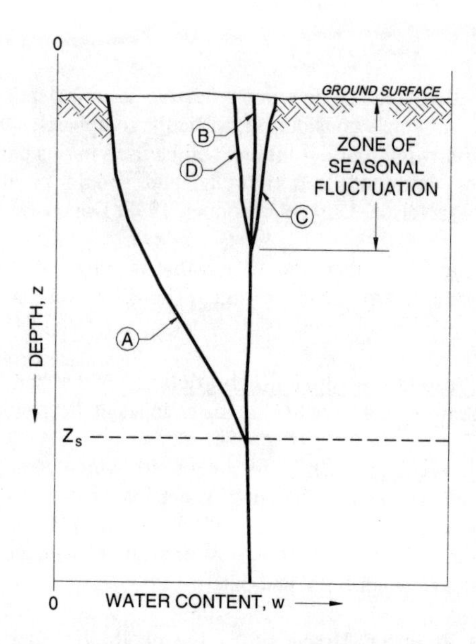

Figure 1: Idealized Water Content Profile

One obvious example of a surface cover is the placement of a slab-on-grade. Another example of a cover is an irrigated lawn, where water is supplied at a rate equal to the evapo-transpiration needs of the grass. Although some water content fluctuations may occur in the upper 150 to 300 mm. (6 to 12 inches) of the soil profile depending on the type of grass, below this surficial zone, the effect is to eliminate all water flux across the upper boundary.

However, if the water introduced by irrigation exceeds the evapo-transpiration needs of the grass, as is normally the case in order to avoid stressing the grass, a net excess of water exists at the soil surface. McWhorter and Nelson (1979) developed a methodology to analyze seepage from a surface water source into an unsaturated soil. Although this methodology was developed for an application different than expansive soils, the scientific and engineering principles governing the movement of water in both cases are exactly the same, and this method of analysis will be applicable to expansive soils.

Figure 2, modified from McWhorter and Nelson (1979), shows a water content profile in the zone of soil beneath a surface water source at times t_1 and t_2 ($t_2 > t_1$). The migration of a wetting front that moves downward with time is shown.

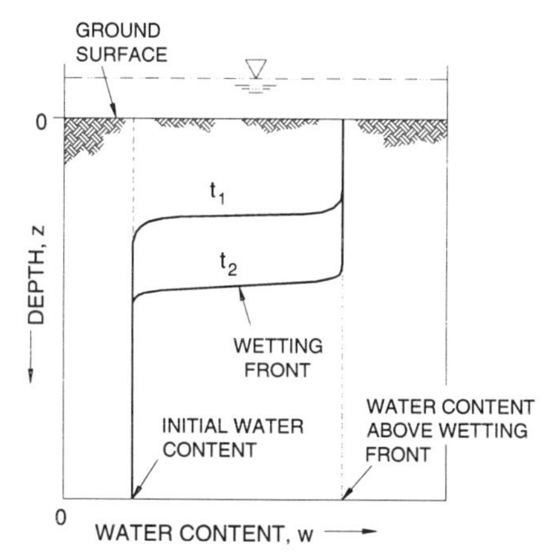

Figure 2: Schematic Distribution of Water Content (After McWhorter and Nelson 1979)

Below the wetting front, the water content is the same as that which existed prior to introduction of the water source. Above the wetting front the water contents are higher, and the soil may or may not be saturated depending on the soil conditions. The depth to the wetting front is the depth of wetting. This wetting front will continue to move downward as long as the total head of the soil above the wetting front is higher than that below the wetting front. For a uniform soil profile the soil suction in the unsaturated zone below the wetting front will be greater than that in the soil above the wetting front where water contents are higher. The difference in soil suction between the wetter and drier zones will result in downward flow of water. The wetting front will continue to migrate downward until an impermeable boundary or a water table is reached. For considerations of expansive soils, even very plastic clays do not represent a truly impermeable boundary. The movement of the wetting front in those soils may be very slow, but even small increases in water content may cause significant amounts of heave.

Figure 3 shows some actual measured data from the Colorado State University field test site (Nelson et al., 1994). The similarity between Figs. 1 and 3 is evident. The slight difference between the two water content profiles for 365 days and 596 days near the ground surface in Fig. 3 shows the effects of temperature.

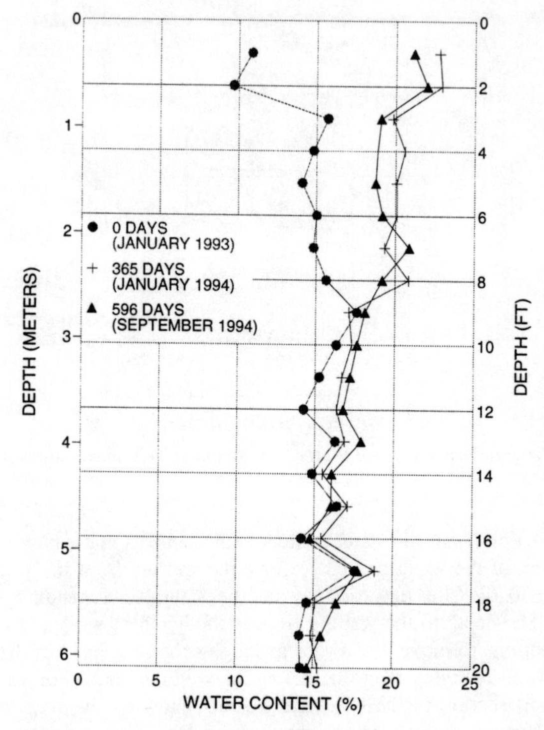

Figure 3: Water Content Profile from CSU Site (Nelson, et al., 1994)

Figure 4 shows actual measured data from a residential site in the Front Range area of Colorado. Water content profiles are shown both before construction and approximately three years after construction of the house. The presence of a wetting front that is progressing downward is evident above a depth of about 4 ½ m. (15 feet).

The above discussion relates to a site in which the soil is fairly uniform, or in which the soil exists as a series of uniform strata. It assumes that there are few or no discontinuities or zones in which flow patterns will be concentrated. Such is the case, for example, in many areas of the Front Range area of Colorado, particularly in the Denver Formation. In those areas the beds are nearly flat-lying and consist of deep deposits of claystone or clayshale.

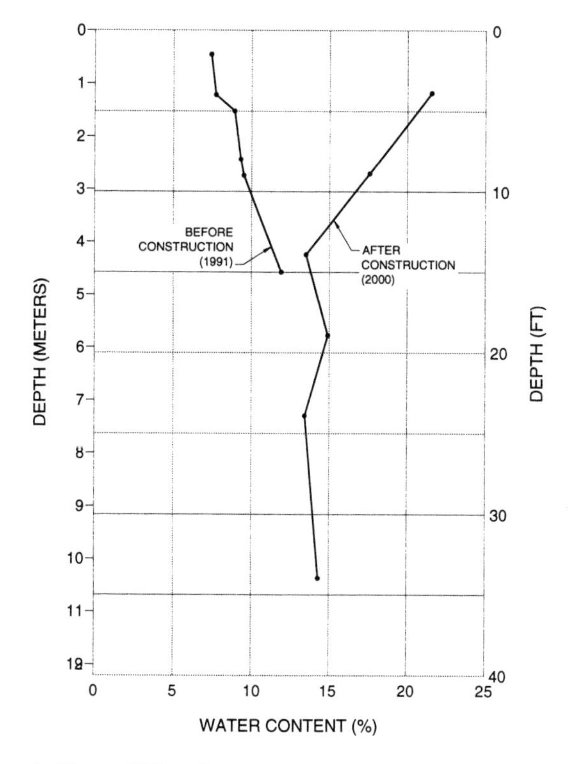

Figure 4: Observed Water Content Profile in Fort Collins, Colorado

In areas closer to the foothills, the bedrock formation is the Pierre formation. At the Expansive Soils Field Test Site on the Colorado State University Foothills Campus, the Pierre Shale dips to the east at about 15° to 20°. This is also true for most of the general area around Ft Collins and Loveland, and in many areas in the western part of the Colorado Springs area. However, there is an area west of Denver and east of the foothills in which the beds of the Pierre formation dip quite steeply to the east and at some locations the dip is almost vertical. The steeply dipping beds pose particular problems and the frequency of distress to structures in these areas is significantly higher than in areas with more flat-lying beds (Thompson, R.W., 1992). The area in which the beds dip at angles greater than 30° is termed the "Designated Dipping Bedrock Area" (DDBA). Special regulations are in effect for building in that area.

In areas where the beds dip, water movement is significantly different than described above. Fig. 5 depicts water movement at a site near Loveland, Colorado. The building at the site was experiencing pier movement in amounts as great as 275 mm.(11 inches) or more. At that location the beds dip to the east at about 20°. Water sources at the site included irrigation and a small pond to the west of the building. Continuous core was taken during drilling and between sampling. Water contents were measured in the laboratory from drive samples taken during the site investigation. In the core, the bedding planes were evident, and along many of them, iron staining could be seen, indicating oxidation due to water traveling along the beds. The clayshale was soft and iron stained for a distance of about 25 to 50 mm. (1 to 2 inches) on either side of the bedding planes. It was postulated that joints and cracks also existed in near vertical directions. These discontinuities provided flow paths between bedding planes. Based on those observations, a conceptual model of water movement as depicted in Fig. 5 was developed.

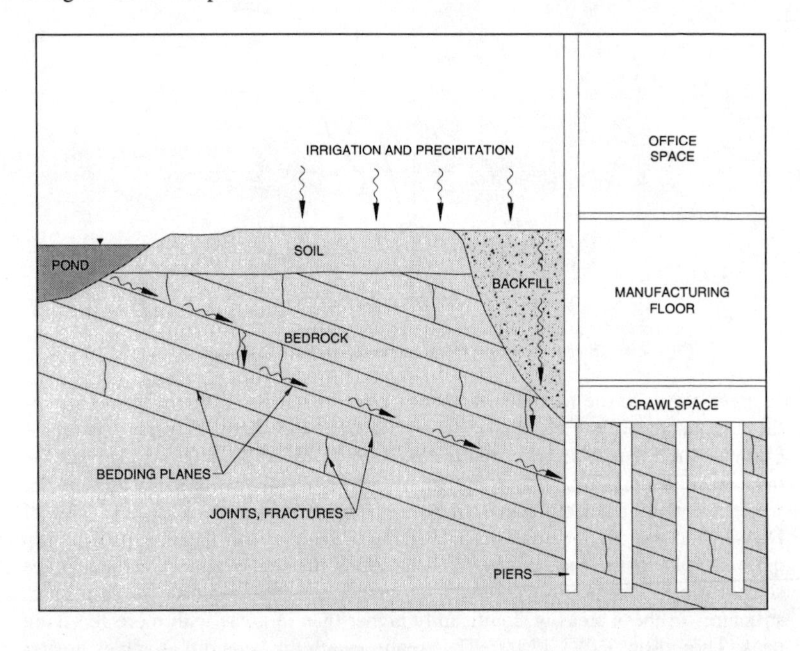

Figure 5: Water Movement in Dipping Bedrock in Loveland, Colorado

Figure 6 shows the water content profile for pre-construction conditions and during the investigation of the distress at the Loveland site. The discreet points measured on the individual samples indicate that there is a downward progressing wetting front. However, observations of the core as described above show that the water contents are not as uniform as indicated in Fig. 6. In fact, adjacent to the bedding planes the soil is quite wet and water migrates outward from there into the zones between the bedding planes. The water movement had progressed to a depth of about 10 m. (34 feet) at the time of the investigation. As water continues to migrate into deeper bedding planes along the joints and cracks, the apparent "wetting front" will continue to move downward, and the clayshale at deeper depths will increase in water content.

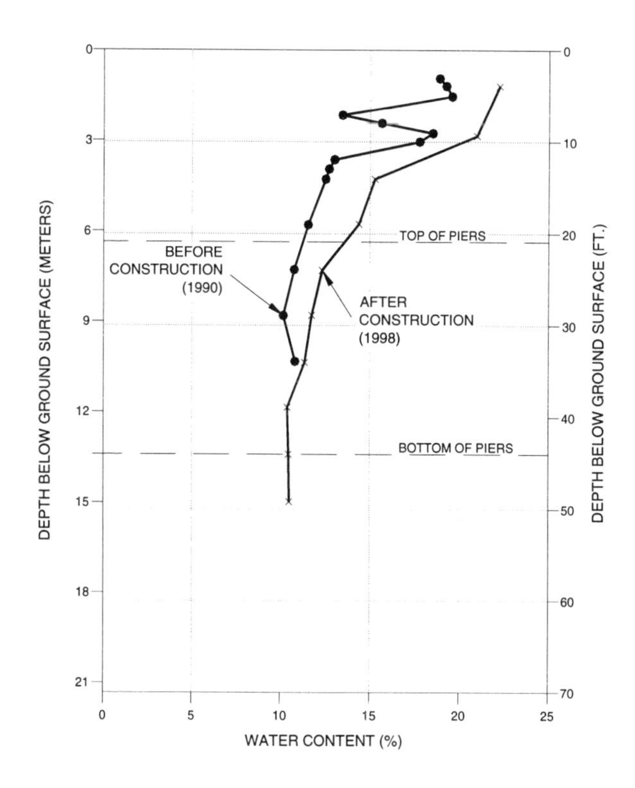

Figure 6: Water Content Profile in Dipping Bedrock in Loveland, Colorado

RATE OF MOVEMENT OF WETTING FRONT.

McWhorter and Nelson (1979) developed a method of analysis to model the movement of the wetting front as shown in Fig. 2. Below the wetting front the pressure head in the soil is dictated by the soil suction, which in turn, is a function of the water content. The analysis is complicated somewhat by the fact that for a small zone across the wetting front the water content, and hence, the suction is varying. McWhorter and Nelson (1979) showed that a good approximation for the pressure head at the wetting front is to assume that the suction is equal to the displacement pressure of the soil. For clay shale this value can be quite high.

According to Darcy's law, for the case where water is continuously ponded to a small depth at the ground surface, the infiltration at the ground surface, Q_i, is,

$$Q_i = K(1 + H_d/Z_w) \qquad (1)$$

where:
- K = the hydraulic conductivity of the clayshale
- H_d = the displacement head of the clayshale(note that a negative water pressure corresponds to a positive value of H_d)
- Z_w = the depth of wetting

The infiltration, Q_i, does not all contribute to downward migration of the wetting front. Some of it is retained in the pore spaces of the soil and increases the water content. This increase in water content is, of course, the factor that causes heave.

As shown in Fig. 4, the water content increases typically from around 15% ± 5% to 25% ± 5% in the Front Range area. Analyses of time rate of wetting front movement, for a typical case where there is a constant source of water at the surface, have shown that to move downward a distance of 10 m. (30 feet) can require 20 to 30 years or more (Durkee, 2000). Many litigation cases come into being 6 to 10 years after construction of the building, and it is not uncommon to see depths of wetting of approximately 6 m. (20 feet) in that time period. As time progresses, the depth of wetting, and hence, distress to the building, will continue.

HEAVE AND PIER MOVEMENT

Figure 7 depicts a pier and grade beam foundation for a structure with a slab-on-grade floor. For purposes of simplicity, the slab is shown at ground level. After some time the wetting front has moved down to the position shown. Again, the wetting front will not be as uniform as shown, and will most likely be deeper outside of the slab. Nevertheless, to demonstrate the effect of the active zone on heave and pier movement, the wetting front movement has been shown as being uniform.

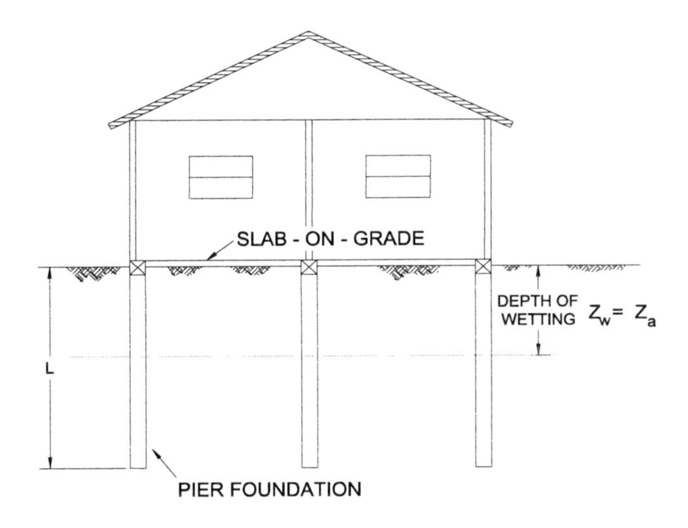

Figure 7: Pier Foundation System with Slab-on-Grade

In Fig. 7, the zone that is contributing to heave is that zone that exists above the wetting front. This, according to the definition presented above, is the **active zone**. Thus, the active zone is varying with time. For purposes of illustration, a hypothetical soil profile consisting of a uniform soil exhibiting 5% swell in a consolidation-swell test and a swelling pressure of 290 kPa (6,000 psf) as measured in a constant volume swell test has been considered. Methods of analysis for heave and pier movement prediction followed the procedures presented in Nelson and Miller (1992), and Nelson, Durkee and Bonner (1998). For purposes of illustration, the wetting front movement is assumed to be as shown in Fig. 8. This is representative of observations of actual cases in the Front Range area.

The maximum predicted heave for this profile is 370 mm. (14.5 inches). The depth of potential heave, in accordance with the definition presented above, is 10 m. (34 feet). Thus the maximum depth of the active zone would be 10 m. (34 feet) and according to Fig. 8, it would require approximately 18 years to reach this depth. For a soil with a greater heave potential, the depth of potential heave would be greater and a longer time would be required for the wetting front to reach the full depth of potential heave.

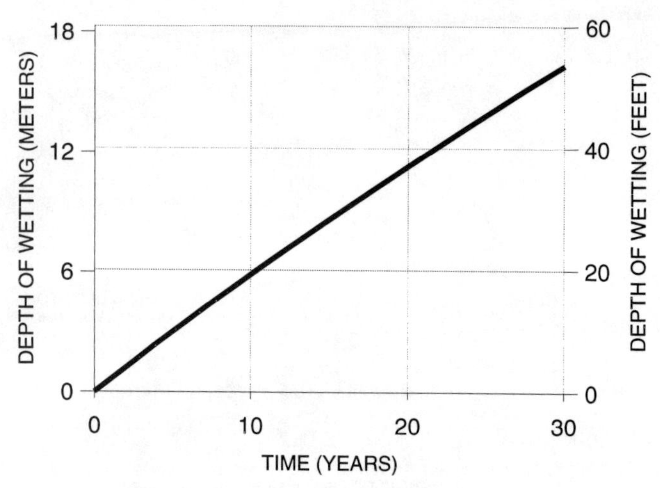

Figure 8: Hypothetical Wetting Front Movement

Figure 9 shows normalized heave of the ground surface, and the slab-on-grade, in the form of ρ_t/ρ_{max}, where ρ_t is the heave at any time and ρ_{max} is the maximum heave that will occur, as a function of normalized depth of active zone, Z_a/Z_p, where Z_a is the depth of the active zone, and Z_p is the depth of potential heave. In this case the depth of the active zone is the same as the depth of wetting. It is interesting to note that this relationship does not depend on swelling potential, and is essentially the same for all soils as long as the profile is uniform.

A primary factor governing pier movement is the ratio of pier length, L, to depth of active zone, Z_a (L/Z_a). Because the depth of the active zone is equal to the depth of wetting, which varies with time, as shown in Fig. 7, the ratio L/Z_a, and hence, pier movement, will also vary with time. For a straight shaft pier, as long as L/Z_a is greater than about 2, the pier movement will be very small (Nelson and Miller, 1992). Thus, pier movement does not begin until the active zone has reached a depth of about half of the pier length. The time of pier movement can be accelerated if deep-seated heave can occur due to groundwater access at greater depths. In this case, however, movement of the wetting front is assumed to progress downward only from the surface. (Reed, 1985)

Predicted pier movement and slab heave were computed for the idealized soil profile described above. Both 6 m. and 11 m. (20 ft and 35 ft) long piers were considered.

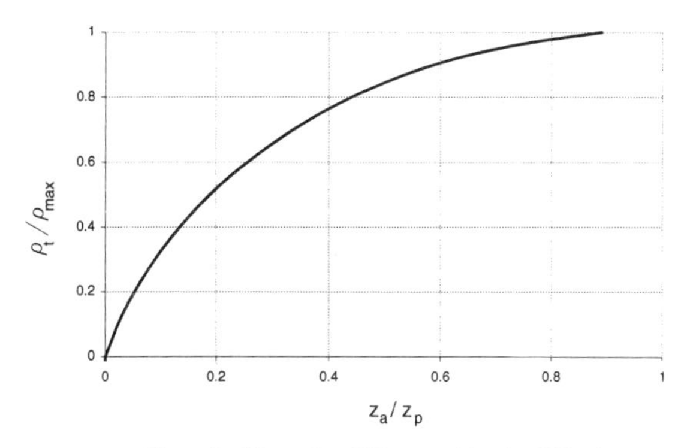

Figure 9: Normalized Heave vs Depth of Wetting

Figure 10 shows slab and pier heave as a function of time for the rate of wetting front movement shown in Fig. 8. Data for only the first ten years after development of a wetting front are shown. It is particularly interesting to note that, whereas slab movement begins almost immediately after a wetting front has begun to develop, pier movement does not begin until several years later. Furthermore, the length of the pier influences both the time at which pier movement begins and the amount of pier movement.

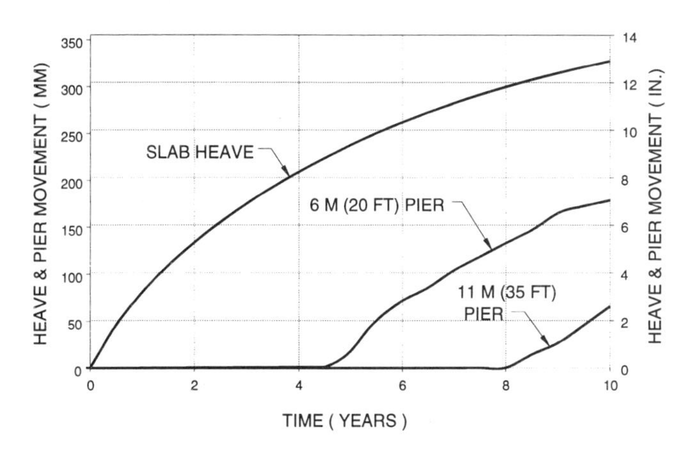

Figure 10: Heave and Pier Movement for Hypothetical Site, First Ten Years

DISCUSSION AND CONCLUSIONS

The definition of active zone has taken on various different meanings as a result of different usage of the term by different engineers in different situations. It is believed that the definition presented above is the most correct, and consistent with general usage of the term. The definitions of the terms depth of wetting, zone of seasonal moisture fluctuation, and depth of potential heave help to differentiate between different historical usages of the active zone.

Water movement in soils is frequently considered to be a uniform migration of a wetting front within a soil mass. However, discontinuities in the soil such as bedding planes, cracks, and fissures have a significant influence on water movement. As water moves within the soil, the active zone changes with time, and movement of structures constructed on and in the soil profile varies with time. It has been shown that whereas movement of a slab-on-grade will commence almost immediately after a wetting front develops, pier movement will be delayed for several years, depending on the length of the pier.

The depth of the wetting front is a very important factor in the design of a foundation system for expansive soils. Unfortunately, it is common practice in the Front Range area for practicing engineers to assume a depth of wetting of only 2 to 3 m. (7 to 10 feet) below basement slabs-on-grade (CAGE, 1996). With the observation that wetting fronts can migrate deeper, the assumed depth of wetting has increased somewhat (CAGE, 1999; McOmber and Thompson, 2000). However, common practice continues to be to assume some finite depth of wetting unrelated to the depth of potential heave.

There is no sound rationale behind assuming that the depth of wetting will stop at some assumed depth. Instead, a more prudent assumption is to assume that, in time, the depth of wetting will extend throughout the entire depth of potential heave. This in fact, is realistic and conservative. Unfortunately, the profession's database of observations does not extend over a sufficiently long time span to confirm this. Usually, litigation occurs within the first ten years after construction. Unfortunately, most of the data related to constructed facilities is collected in the course of litigation, and little long-term data is available. However, data collected during geotechnical investigations related to groundwater contamination and contaminant migration have shown that wetting front migration can extend for depths much greater than 6 m. (20 feet). Data such as these cannot be ignored when determining the depth of wetting for foundation design.

REFERENCES:

Colorado Association of Geotechnical Engineers (CAGE) (1996), Guideline for Slab Performance Risk Evaluation and Residential Basement Floor System Recommendations (Denver Metropolitan Area) *plus* Guideline Commentary, Available to members only, not available to the public.

Colorado Association of Geotechnical Engineers (CAGE) (1999), Drilled Pier Design Criteria for Lightly Loaded Structures in the Denver Metropolitan Area, Available to members only, not available to the public.

Durkee, D. B. (2000), "Active Zone and Edge Moisture Variation Distance in Expansive Soils", Dissertation submitted in Partial Requirement for the PhD Degree, Colorado State University, Fort Collins Colorado.

McOmber, R.M. and Thompson, R.W. (2000), "Verification of Depth of Wetting for Potential Heave Calculations", Advances in Unsaturated Geotechnics, ed. Shackelford et al., ASCE Geotechnical Special Publication No. 99 pp. 409-422.

McWhorter, D. B. and Nelson, J. D. (1979), "Unsaturated Flow Beneath Tailings Impoundments," J. GeotechEngDiv, ASCE, Vol. 105, No. GT11, November, pp. 1317-1334.

Nelson and Miller, (1992) Expansive Soils: Problems and Practice in Foundation and Pavement Engineering, John Wiley and Sons, New York.

Nelson, J.D., Durkee, D.B., Reichler, J.D., and Miller, D.J. (1994), "Moisture Movement and Heave Beneath Simulated Foundation Slabs on Expansive Soils", prepared for US Army Corps of Engrs. by Civil Engr. Dept., Colorado State University, 136 p.

Nelson, John D., Dean B. Durkee and James P. Bonner, (1998), "Prediction of Free Field Heave Using Oedometer Test Data". Proceedings of the 46th Annual Geotechnical Engineering Conference, University of Minnesota, St. Paul, Minnesota, February 20, 17 p.

Reed, R.F. (1985), "Foundation Performance in an Expansive Clay", Proceedings of the 38th Canadian Geotechnical Conference, Theory and Practice in Foundation Engineering, Edmonton, Alberta, September. (5 ppg.)

Thompson, R.W. (1992), "Performance of Foundations in Steeply Dipping Claystone", Proc. 7th Intl. Conf. Expansive Soils, Dallas, Texas (USA), 84-88.

TREE ROOT INFLUENCE ON SOIL-STRUCTURE INTERACTION IN EXPANSIVE CLAY SOILS

John T. Bryant, Derek V. Morris, Sean P. Sweeney, Michael D. Gehrig, J. Derick Mathis [1]

ABSTRACT

Tree roots have been documented to withdraw volumes of water from expansive clay soils causing slab distress under certain circumstances. However, patterns of distress, areas of influence and potential distress magnitudes have not been systematically considered from a soil-structure interaction perspective. This Paper describes research into the soil-structure interaction between trees and heavy vegetation and at-grade structures including pavements and slab-on-grade foundation systems. Several areas of distressed pavements and slab-on-grade foundation systems in the North Texas area have been documented, with estimates and measurements of soil properties, tree geometry and proximity to the structure, tree concentration, relative elevation surveys and distress patterns to confirm the influence pattern of the root systems. Conclusions indicate concentrations of trees tend to have the most significant effects with influence patterns occurring radially to approximately the tree drip line. Confirmation of the horizontal and vertical influence pattern is provided by Geo-Electrical Moisture Material Imaging Resistivity (GMMIR) surveys using soil borings to help characterize the influence pattern of trees. The GMMIR plots indicate influences between about 1.2 and 3 meters in depth. Root influence models were constructed using unsaturated soil mechanics theory and the VOLFLO algorithm. Analysis of these models indicates that the influence of tree roots is a function of the soil type and that the influence at the extreme perimeter of the structure is as significant as the penetration of the tree roots significantly beneath the foundation with differential downward movements on the order of 127 to 152 millimeters (5 to 6 inches) at the perimeter moving from a wet condition (total soil suction pG = 4.3 (pF = 3.3)) to a desiccated tree condition (total soil suction pG = 5.6 (pF=4.6)) reducing to approximately 5.1 centimeters (2 inches) or less when the tree roots are at least 0.6 meters (2 feet) from the structure perimeter.

[1] Bryant Consultants, Inc., 4393 Westgrove Drive, Addison, Texas 75001. Phone 972-713-9109, Email: jbryant@geoneering.com, Website: geoneering.com

INTRODUCTION

Plotnik (2000) indicates that the shape of a root mass is as variable as the crown shape. Sometimes it mirrors the crown, spreading to the drip line or margin of outermost leaves where the water drips off. In contrast, the roots may form a tight clump where water collects under a narrow tree. The roots must supply the water to the leaves where the photo-chemical reactions are occurring. Plotnik also indicates that the roots absorb and elongate only at the root tips, which push their protective lubricated caps through the path of least resistance and the most moisture. Legget and Crawford (1965) indicate that about 10 percent of the wood mass of a tree is to be found under the ground surface in the form of roots.

However, the critical question is not about the exact root ball dimensions, but about the effect of the roots on the soil and structures in proximity to the tree itself. Do the roots mirror the crown or canopy? Do the roots extend past this canopy? Quests to understand these phenomena have been attempted to some degree by others; however, coupling of the distress pattern and the soil movement mechanism has not been extensively researched. To answer the questions and to understand the phenomenon of Soil-Vegetation-Structure interaction (SVS) will require direct measurements of pavement or structure distress, tree dimensions and observations coupled with using inductive reasoning, geotechnical and geophysical exploration and the theoretical basis of unsaturated soil mechanics. Systematic investigations using this panoply of tools can lead to some clearer understanding of the (SVS) interaction system.

As early as 1947, the British Building Research Station had published information that stressed the importance of careful building near trees and shrubs. Legget and Crawford (1965) identify trees as the culprits in many cases of foundation problems in Ottawa, Canada, but then give some preventative measures to control them including consideration of the soil types, considerations for natural trees, and considerations for new tree plantings. Legget and Crawford (1965) indicate that consideration of new plantings should include distances from the structure equal to the mature tree height. Obviously, some of the proscriptions in the Legget and Crawford (1965) research may not be practical when considering the overall site and economy of the SVS system.

Hamilton (1977) noted that shallow foundation problems associated with shrinkable clay soils are almost exclusively due to the drying influence of tree roots in the Leda Clay found in the St. Lawrence and Ottawa Valleys in Canada. Hamilton also noted that many tree-lined streets in several cities in Western Canada possess the tell-tale bowl-shaped depressions in sidewalks, roadways and landscaping, and foundation distortions give evidence of the progressive shrinkage settlements often of 300 millimeters (12 inches) or more. He also noted that removal of heavy tree growth from a building site can result in foundation heaves.

Biddle (1998) noted that the interaction of the tree and the soil involves water, and it is the constant movement of water within the system, which produces the dynamic condition. These movements of water and their interaction with the soil and the tree cause the soil-structure interaction issues that can potentially cause movement and possibly damage to a structure. Indications are that the higher the plasticity of the clay, which are directly correlated to the higher the amount of clay fraction and plasticity index, the more severe the movements and interactions become. In other words, expansive clay soils that exist in unsaturated states are the most susceptible to this phenomenon.

RELATIONSHIP BETWEEN TREE TYPE AND DAMAGE SEVERITY

Table 1 summarizes previously published water usage and damage data for various tree species and ranks them from several different researchers over several decades. The degree of severity is most closely related to the water demand. Higher water demand would indicate a higher degree of severity due to the desiccating effects of the tree roots and their extraction of water causing changes in the effective stress and ultimately reducing the volume of the expansive clay soils.

Table 1. The degree of severity of the various species of trees.

Rank as a function of Damage/Water Demand	**Ward (1947) Britain**	**Hammer & Thompson (1966) USA**	**Biddle (1978) Britain**	**Driscoll (1984) Britain**
1	Poplar	Chinese Elm	Poplar	Oak
2	Adler	American Elm	Poplar	Poplar
3	Aspen	Poplar	Monterey Cypress	Lime
4	Willow	Willow	Whitebeam	Ash
5	Elm	Oak	Hawthorn	Plane
6	Sycamore	Maple	Oak	Willow
7	Birch	Ash	Elm	Elm
8	Ash	Sycamore	Ash	Hawthorn
9	Beech	Norway Maple	Lime	Maple/Sycamore
10	Oak	Hawthorn	Plane	Cherry/Plum
11	Larch	--	Sycamore	Beech
12	Spruce	--	Birch	Birch
13	Fir	--	Beech	Whitebeam / Rowan
14	Pine	--	Mulberry	*Cupressus Macro...*

NOTE: The Relative Average Rank (RAR) is computed by taking the ranks in Table 1 and dividing by the number of occurrences in Table 1 for the most frequently occurring species.

Table 2. Common List Trees Relative Average Rank From Table 1.

New Rank	Relative Average Rank (RAR)	Species
1	1.8	Poplar
2	4.4	Elm
3	4.7	Willow
4	5.5	Oak
5	6.8	Ash
6	8.5	Sycamore
7	10.3	Birch
8	11.0	Beech

The above analysis and comparisons reveal some general trends, which indicate that certain species of trees potentially have a more severe effect on structures than other species. Further, climatic influences including long drought and rain periods are also a factor in the effective stress state of the soils and can exacerbate the influence of the trees or heavy vegetation. As a result, the research also indicates that substantial variation exists between individual trees of the same species, caused either by soil site characteristics or due to genetic variation between individuals and climatic influence. These individual and site differences can over-shadow and blur the lines between the differences between various species. In other words, a large Birch near the structure may cause as much damage as a much smaller Poplar tree near the structure. This result indicates the validity of the concept of non-uniqueness and variability of tree effects on structures. The results of the Table 2 ranking indicates that common oak trees have significant water demand. Oak species are common in the DFW area.

PATTERNS OF DISTRESS

Biddle (1998) has done extensive research plotting moisture contents and tree patterns to understand the influence of the tree roots in Great Britain and their relationship to expansive clay soils. Further, the presence of expansive clay soils has been documented in many parts of the United States. These expansive clay soils are present in Texas, and are especially concentrated in the rocks of the Cretaceous and younger age strata. Wiggins (1976) has mapped the distribution of expansive clay soils in the United States and their relative activity.

The ranking analysis demonstrated above assumes trees acting individually. But reason would dictate that synergistic influences and possible modification of distress patterns would occur with groups of trees versus individual trees and as a function of the soil type and its expansive characteristics. To better understand the patterns of tree influence, we have evaluated the areal and depth influence of tree roots by two independent methods: 1) visual observation of distress and 2) indirect electrical resistivity geophysical measurements and soil resistivity measurements.

Visual Observations and Measurements of Distress

The tree sites are situated in several different expansive soil types and formations including soils derived from the Woodbine formation and Austin Chalk formation of Upper Cretaceous geologic age. Most of the trees appeared to be oak species with live oaks found some locations; however, the exact type of tree correlated to the damage at the test sites was not an objective of this investigation

Four sites were chosen across the DFW area due to their noticeable deflection and patterns of distress in both the concrete and asphalt pavements. Figures 2 through 5 provide USDA Soil Conservation data, aerial photographs and distress patterns established for the four sites. Sites 1 and 4 are in close proximity in Richardson, Texas and are situated in the highly expansive Houston Black clay derived from residual weathering of the Austin Chalk Formation. Site 2 is in Addison, Texas and is situated in highly expansive Dalco Urban Land Complex soil, which is typically not very thick in this location and derived from the Austin Chalk Formation. Site 3 is located in the Woodbine formation and juxtaposed between the expansive Navo Clay loam and the less expansive Justin fine sandy loam soils. Photographs of the distress patterns of the pavements at each site are found in Figures 6 and 7. Actual soil testing from several test sites confirmed the presence of expansive clay soils at these locations.

Analysis and review of Figures 2 through 7 indicate several general patterns. These patterns indicate the relative proximity of distress and the geometry of the distress are related to the tree geometry. That is, a more linear band of trees produces a more linear distress pattern as shown in Figures 2 and 6 at Site 1 and especially in Figures 4 and 7 at Site 2. The skewed nature and linear trend of the distress is directly correlated to the tree geometry. Further, the distress patterns in the pavement at Sites 1 and 2 indicate that for the more linear groupings of trees, the distress pattern is more closely correlated to the drip line of the tree as shown in Figures 2 and 4, respectively. In contrast, it appears that the influence pattern of individual trees and small clusters of trees, not oriented in a linear pattern tend to extend past the drip line of the tree more extensively as shown in Figures 3 and 5.

Geophysical GMMIR Evidence for Tree Influence Patterns

The USDA Forest Products Laboratory (1987) has established that the electrical resistivity or conductivity of wood varies greatly with moisture content, especially below the fiber saturation. As the moisture content of the wood increases from near zero to fiber saturation, the electrical resistivity decreases by 10^{10} to 10^{13} times. Therefore, substantial resistivity contrasts between the root system and the surrounding soils are possible.

Geophysical exploration and testing using proven and well-researched electrical resistivity techniques modified and patented by Bryant (1997-2001) to understand soil-structure interaction phenomena and mechanisms called Geo-Electrical Moisture Material Imaging Resistivity (GMMIR) were used to help identify the root influence zone areas. The electrical resistivity of wood, soil and water are typically different. Their resistivity contrasts are functions of the soil type, connate water electrolytic fluid composition and the moisture content of the soil and wood.

The GMMIR technique's efficacy in detecting tree root influence is primarily a function of the resistivity contrast of the soils. In other words, a sufficient resistivity contrast between the soil and the tree root must exist for the influence to be detected by the GMMIR technique. In some instances, wood possesses electrical resistivities that are greater than that of the surrounding soil; however, because the moisture content of the root zones can vary as a function of moisture content, tree roots can also show to have higher electrical resistivity than the surrounding soil. These values and contrasts may also be influenced by the backfill and soils around the tree root ball. Comparison of Figures 8 and 9 show the difference in the root influence zones using the GMMIR technique.

Figures 8 and 9 provide GMMIR plots of several areas in the DFW area where the tree root patterns show distinct resistivity contrasts with the surrounding soil. The effect of multiple tree interactions is shown especially in Figure 8B where a row of trees has influenced the upper 1 to 2 meters of the soil profile.

Figure 9 shows the effects of tree root influence near a residential structure located in the sandy soils of the Woodbine formation of Upper Cretaceous geologic age. As is shown in this GMMIR image, the electrical resistivity of the root zone is lower than that of the surrounding soil, indicating the uptake of moisture from the surrounding soil. This image also shows the influence at this site of a buried subterranean stream channel that was also causing volumetric changes of the soil and differential movements of the soil towards the front of the structure.

In contrast, Figure 8B shows the influence of a group of trees along the side of a residential structure in the same Woodbine formation. The Woodbine soils consisted of clays, with lower natural electrical resistivity values, in contrast to the compounding desiccating effect of the tree group causing the soil resistivities to become much higher. Figure 8A shows the symmetric influence of tree roots in the Eagle Ford shale formation where the surrounding root mass has a higher electrical resistivity value than the highly expansive residual clay of the Eagle Ford formation.

Based upon the GMMIR studies, it would appear the depth of influence of most of the tree roots would be on the order of 1.2 meters (4 feet) to about 3 meters (10 feet) with an average of about 2 meters (6.5 feet).

MOVEMENT MAGNITUDE PREDICTIONS DUE TO ROOT DESICCATION

Expansive Clay Soils and Unsaturated Soil Mechanics Theory

Fredlund and Rahardjo (1993) and Lytton (1994) established the basis for unsaturated soil mechanics theory and the prediction of movements based upon this theory. The influence of various site conditions including vegetation on the design of engineering structures has also been established by Lytton (1997). Bryant (1997) has established that the Dallas/Fort Worth Metroplex (DFW) area lies within a climatic area where extreme variations in soil suction can occur.

Further, the DFW area lies within the Upper Cretaceous and Lower Cretaceous sedimentary rock and Quaternary aged alluvial deposits, which weather to expansive clay soils. The sedimentary rock strata dip gradually towards the south and southeast and increase in age from east to west. The sedimentary strata present from approximately the east boundary to the west boundary of the area consist of in order: Ozan-lower Taylor Marl, Austin Chalk Limestone, Eagle Ford Shale, Woodbine Formation including sands, clays shales and sandstones, Main Street/Paw Paw Limestone and the Paluxy formation consisting predominantly of sands and sandstone strata. The interbedded sedimentary rock formations are typically dissected by the Trinity River and its tributaries, which deposited Quaternary aged sands, silts, sandy clays and gravels along the present and ancient channels and flood plains of these rivers an creeks. Our study involves the geologic facies associated with the Austin Chalk, Eagle Ford Shale and Woodbine formations. Recent or Quaternary geomorphic influences from alluvial and colluvial action have also reshaped and influenced many of the older formations.

According to the Geologic Atlas of Texas (1987) and our experience with the geologic formations in this area, the Austin Chalk formation typically consists of surficial dark brown or gray clay underlain by yellowish brown calcareous clays which grade into tan and gray limestone. The Eagle Ford formation generally consist of the surficial dark yellow and grayish brown residual clays grading to the yellowish brown to olive gray shaley clay. Harder gray shale is typically found at depth. The Woodbine formation typically consists of interbedded soil and rock strata including expansive clays and shaley clays, with clayey sands, sandy clays, sandstones, and shales. Each of these formations possess substantial volumes of unsaturated expansive clay soils.

VOLFLO Analysis for Typical DFW Soil Types

VOLFLO (1989) is a computer program that allows rapid analysis of unsaturated soil volume change and flow calculations in expansive clay soils based upon the theoretical considerations of unsaturated soils as developed by Lytton and described in various publications by Lytton. The primary assumption in the VOLFLO algorithm is that horizontal flow of moisture is causing the volume change, while the vertical component of the flow is restricted due to the structure covering such as the pavement or slab foundation system. This is the most common situation for the at-grade structures analyzed in this study.

Figures 2 through 5 and the USDA Soil Conservation's Soil Survey information were used to help establish the typical soil plasticity properties for the soils derived from these typical geologic formations encountered in the DFW area. Further, Table 3 provides the geotechnical and physical input parameters used in the VOLFLO analysis.

We have assumed for our analysis that the soils are desiccating, that is, they are moving from a wetter state pG=4.3 (pF=3.3) to a drier state pG=5.6 (pF=4.6) as a result of the withdrawal of moisture from the soils due to vegetation. We have further assumed a relatively long edge moisture variation distance of 3 meters (10 feet) and an active zone of about 3 meters (10 feet), which are reasonable.

Table 3. Typical DFW Soil-Vegetation-Structure (SVS) Interaction Models

VARIABLE	SVS MODEL 1 FAT CLAY	SVS MODEL 2 LEAN CLAY	SVS MODEL 3 SANDY CLAY
e_m, m e_m (ft)	3.0 (10)	3.0 (10)	3.0 (10)
vcgn	0.163	0.076	0.061
% clay	70	50	30
k	0.33	0.33	0.33
f	0.5	0.5	0.5
Initial Soil Suction pG (pF)	4.3 (3.3)	4.3 (3.3)	4.3 (3.3)
$D_{c\text{-}pF}$, m (ft)	3.0 (10)	3.0 (10)	3.0 (10)
Final Soil Suction pG_{tree} pF_{tree}	5.6 (4.6)	5.6 (4.6)	5.6 (4.6)

Definitions:

em = edge moisture variation distance

vcgn = volume change guide number

y_m = vertical movement due to moisture flux

k = lateral earth pressure coefficient

f = vertical volume change coefficient

$D_{c\text{-}pF}$ = depth to constant soil suction

Figure 10 presents the results of the VOLFLO analysis. The VOFLO analysis indicates that the most significant effects of the trees occur near the tips. This comports with the research by Plotnick (2000), who indicates that the tree roots absorb and elongate only at the root tips. Further, Figure 10 also indicates the dramatic influence that trees have in higher expansive potential CH clay as defined by the Unified Classification System (USCS) with over 127 millimeters of movement (5 inches) potential as compared to less than 51 millimeters (2 inches) for the lower plasticity CL and SC (USCS) clays. Further, the research indicates that the influence of the tree roots is substantially reduced at about 0.6 meters (2 feet) from the structure decaying rapidly away from the structure.

CONCLUSIONS

1. Based upon the relative average rank analysis, the most influential trees are in order: 1) Poplar, 2) Elm, 3) Willow, 4) Oak and 5) Ash.
2. Based upon visual and geophysical measurements and observations, groups of trees tend to modify the structure and pavement distress effects with linear tree patterns producing more linear distress patterns.
3. Distress patterns tend to follow the drip line of the tree and extend out radially from the tree more substantially for individual trees or clusters versus linear groupings.
4. GMMIR techniques can be used to understand the horizontal and vertical patterns of tree root influence.
5. GMMIR techniques indicate vertical influence patterns on the order of 1.2 meters (4 feet) to about 3 meters (10 feet) with an average of about 2 meters (6.5 feet).
6. Soil-structure interactions indicate downward movement near the drip line of the tree with more substantial movements near the root tip with changes of the moisture state from lower total suction to higher total soil suction as noted by empirical observation and by unsaturated soil mechanics predictions.
7. The research indicates that a change in the total suction must occur to cause volume change, and thus in the presence of a constant moisture source or constant suction regime, no volume change due to the vegetation influence will occur.
8. Differential movement magnitudes are a function of the soil type with highly expansive clay or CH (USCS) soils having the most substantial movements of approximately 127 millimeters (5 inches) to 152 millimeters (6 inches), with lower plasticity lean clay and sandy clay (CL and SC, USCS) soils having much less substantial movement magnitudes of about 51 millimeters (2 inches) or less.
9. Significant reduction in movement is noted at relatively short distances from the edge of the tree root tips as modeled by VOLFLO and by the visual observations at canopy.

REFERENCES CITED

Biddle, P.G. (1979). Tree root damage to buildings-an arborculturist's experience. Arbor. J. 3, (6), 397-412.

Biddle, P.G., (1998). Tree Root Damage to Buildings, Volumes 1 and 2. Acorn Press, Swindon, England, Volume 1 376 p, Volume 2 299 p.

Bryant, J.T., (1997-2001). US Patent Application S/N 09/071,577, US Patent Office, Washington, D.C.

Bryant, J.T., (1997). Variations of Soil Suction with Depth in Dallas and Fort Worth, Texas. Applications of Emerging Technologies in Transportation, Transportation Research Record, No. 1615, p 100-104.

Bureau of Economic Geology, (1987). Geologic Atlas of Texas, Dallas Sheet. University of Texas at Austin.

Forest Products Laboratory, (1987). Wood Handbook: Wood as an Engineering Material, USDA, Forest Service, Agricultural Handbook 72, Washington, D.C., 466 p.

Fredlund, D. G. and Rahardjo (1993). Soil Mechanics for Unsaturated Soils. Wiley, New York, 517 p.

Hamilton, J.J. (1977). Foundations on Swelling or Shrinking Sub-soils. Canadian Building Digest (CBD)-184, 1-7.

Hammer, M.J and Thompson, O.B. (1966). Foundation clay shrinkage caused by large trees. J. Soil Mech. Founds. Div. Proc. Am. Soc. Civ. Eng. 92, 1-17.

Legget, R.F. and Crawford, C.B. (1965). Trees and Buildings. Canadian Building Digest (CBD)-62, 1-6.

Lytton, R. L. (1994). Prediction of Movement in Expansive Clays. Vertical and Horizontal Deformation of Foundations and Embankments. Geotechnical Special Publication No. 40. Yeung, A.T. and Filio, G.Y., eds., ASCE, New York, Vol. 2, p 1827-1845.

Lytton, R. L. (1997). Engineering structures in expansive soils. Proceedings of the Third International Symposium on Unstaturated Soils, Rio de Janeiro, Brazil.

Plotnik, Arthur, 2000, The Urban Tree Book. Crown, NY, NY., 430 p.

Soil Conservation Service, (1980). Soil Survey for Dallas County Texas. USDA, Washington, D.C.

Soil Conservation Service, USDA, (1980). Soil Survey for Denton County Texas. USDA, Washington, D.C.

VOLFLO, (1989). Users Manual for Volume Change and Flow Calculations in Expansive Clay Soils. Ray D. Ullrich & Associates, Inc., Austin, Texas.

Wiggins, John H. (1976). Natural Hazards, An Unexpected Building Loss Assessment, Technical Report No. 1246, J.H. Wiggins Company, Redondo Beach, Ca., pp 95-134.

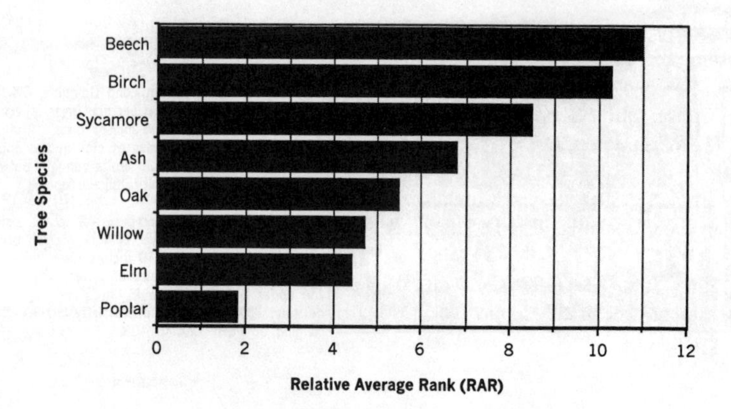

Figure 1. Relative Average Rank From Table 2 According to Species.

Note: A higher rank, i.e. RAR=1 is considered more severe than a lower rank, i.e., 10.

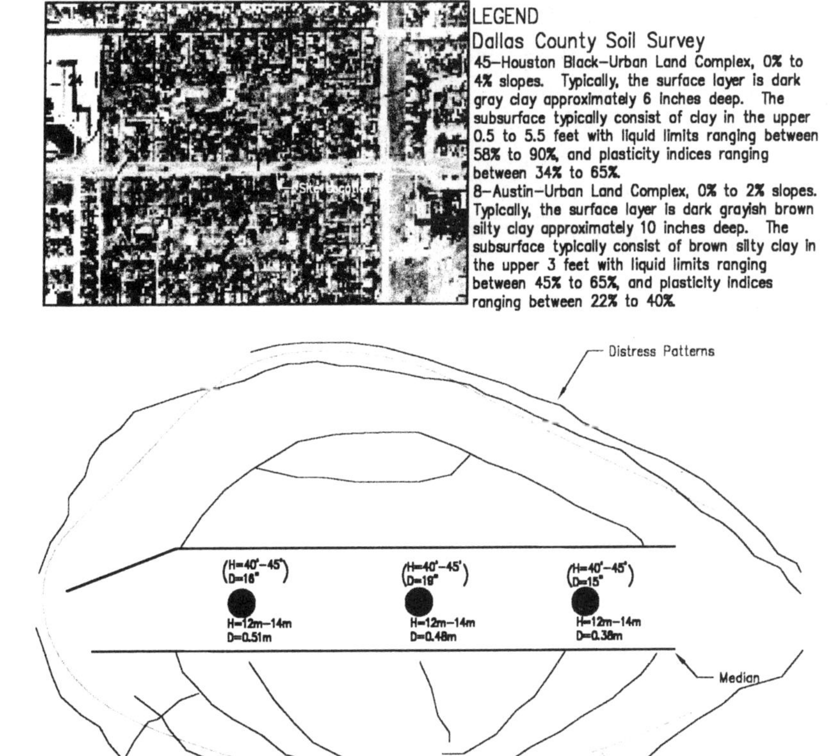

FIGURE 2 – SITE 1: TREE INFLUENCES ON A CONCRETE STREET SECTION IN RICHARDSON, TX

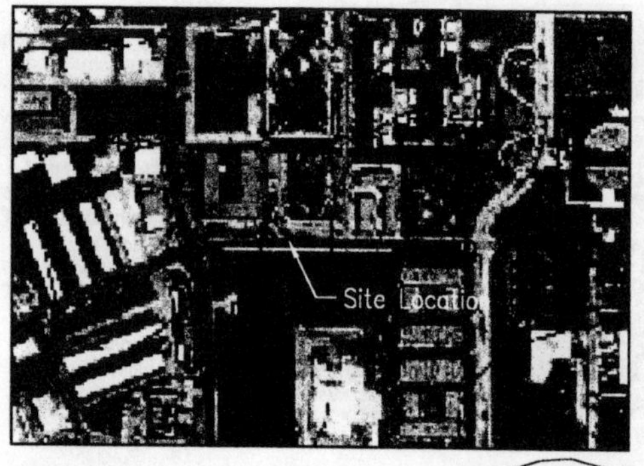

LEGEND
DALLAS COUNTY SOIL SURVEY
24—Dalco—Urban Land Complex, 0% to 3% slopes. Typically, the surface layer is black clay approximately 2 feet deep. The subsurface consist of dark gray clay in the upper 3 feet with liquid limits ranging between 55% to 75% and plasticity indices ranging between 32% to 50%.

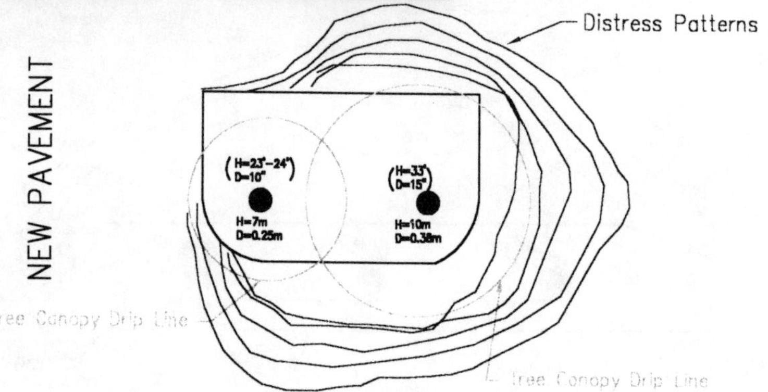

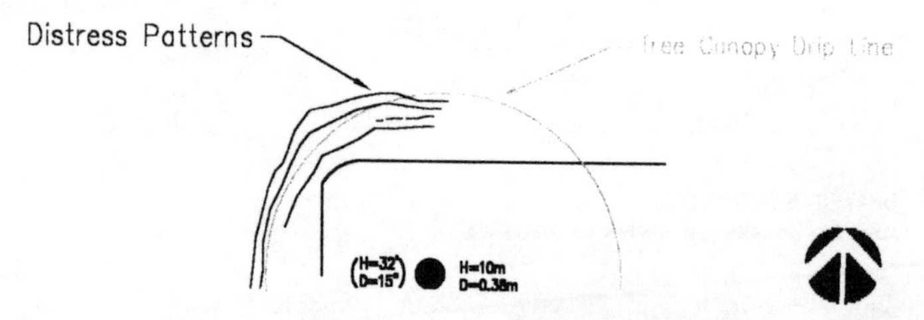

NORTH

DRAWING SCALE: 1:200
Note: Tree dimensions and locations are approximate.

FIGURE 3 – SITE 2: TREE INFLUENCES ON AN ASPHALT SECTION IN ADDISON, TX

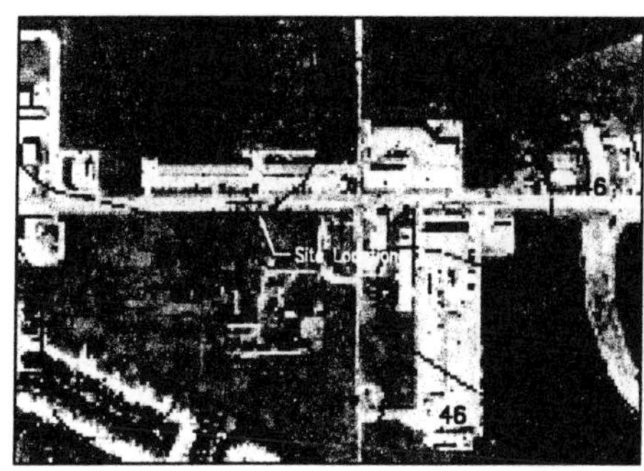

LEGEND

Denton County Soil Survey

46–Justin Fine Sandy Loam, 1% to 3% slopes. Typically, the surface layer is brown fine sandy loam approximately 1 foot deep. The subsurface typically consist of clay loam in the upper 1 to 6 feet with liquid limit ranges between 20% to 52%, and plasticity indices between 8% to 30%.

60–Navo Clay Loam, 1% to 3% slopes. Typically, the surface layer is brown clay loam approximately 5 inches deep. The subsurface typically consist of clay in the upper 0.5 to 6 feet with liquid limit ranges between 25% to 65%, and the plasticity indices between 11% to 45%.

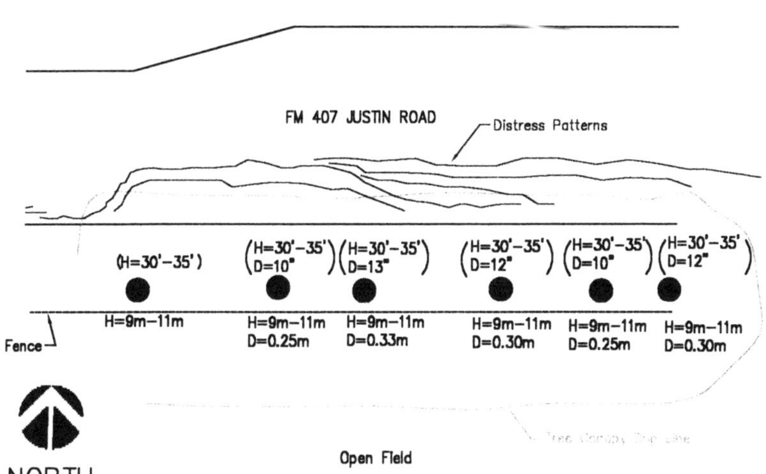

DRAWING SCALE: 1:300

Note: Tree dimensions and locations are approximate.

FIGURE 4 – SITE 3: TREE INFLUENCES ON A CONCRETE SECTION IN LEWISVILLE, TX

LEGEND
Dallas County Soil Survey
45–Houston Black–Urban Land Complex, 0% to 4% slopes.
Typically, the surface layer is dark gray clay approximately 5 inches deep. The subsurface typically consist of clay in the upper 0.5 to 5.5 feet with liquid limits ranging between 58% to 90% and plasticity indices ranging between 34% to 65%.

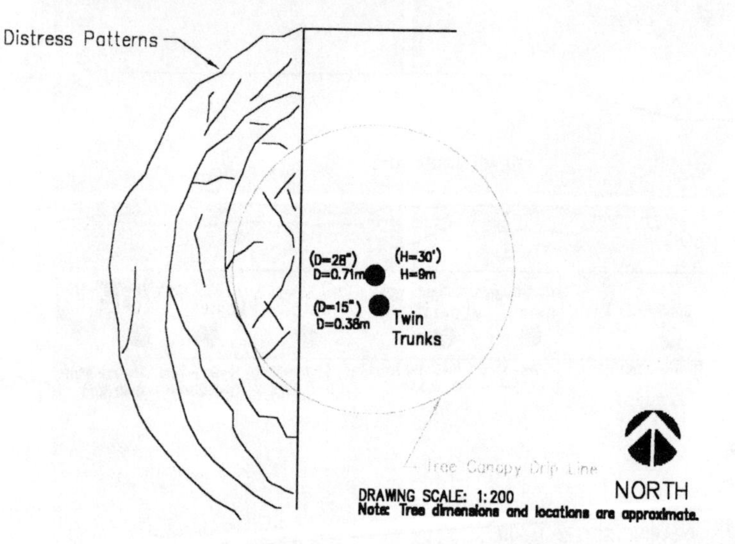

FIGURE 5 – SITE 4: TREE INFLUENCES ON A CONCRETE SECTION IN RICHARDSON, TX

Figure 6: Photographic Documentation of Distress at Sites 1 and 2.

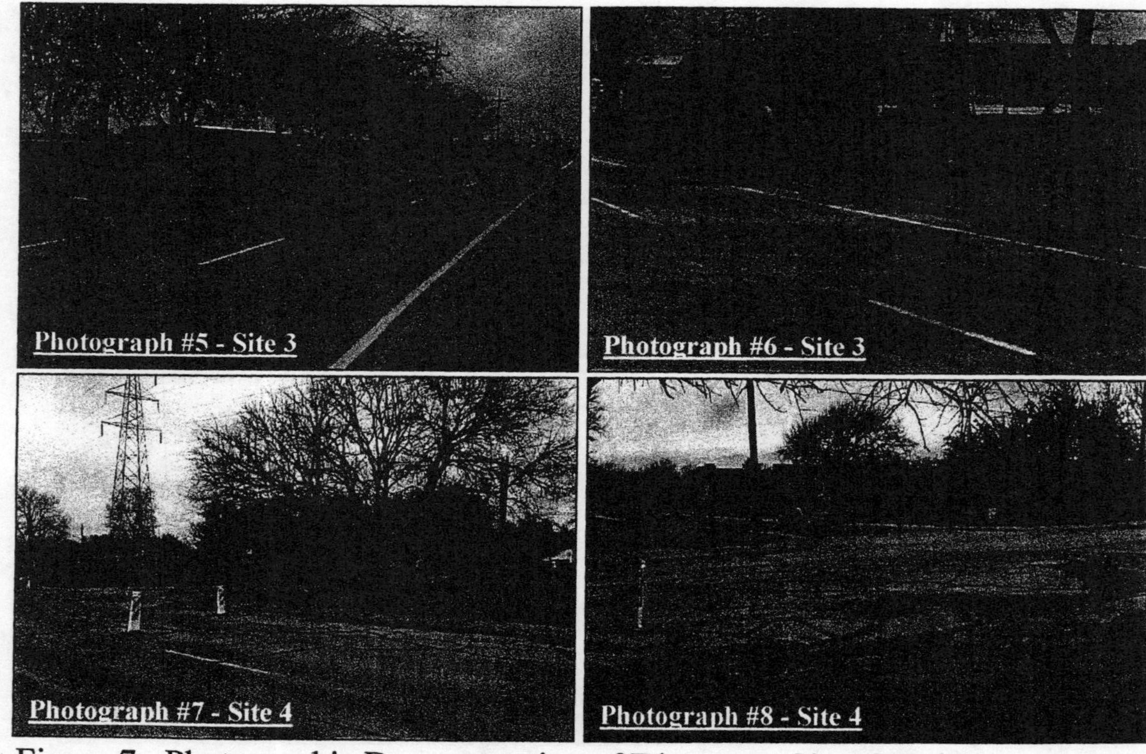

Figure 7: Photographic Documentation of Distress at Sites 3 and 4.

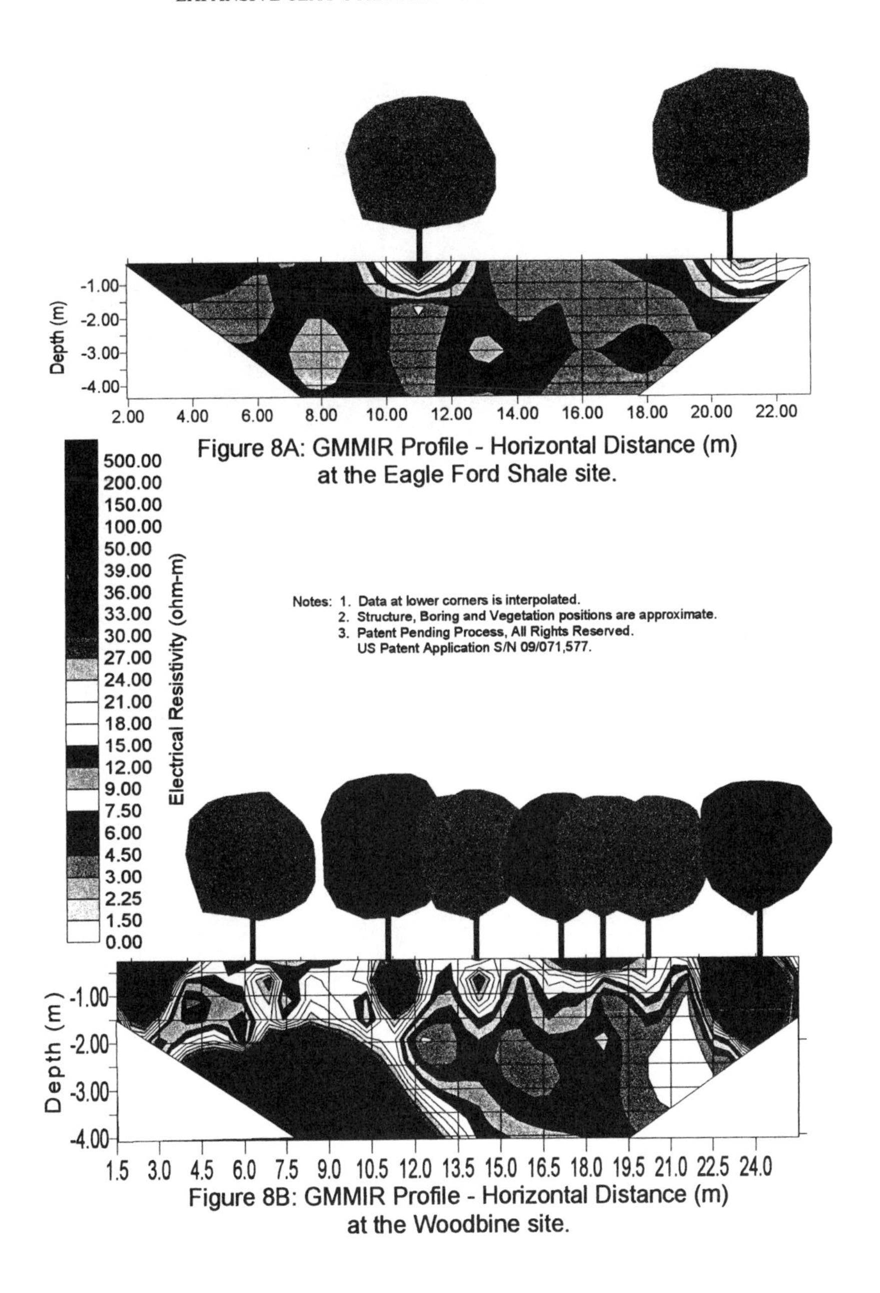

Figure 8A: GMMIR Profile - Horizontal Distance (m) at the Eagle Ford Shale site.

Figure 8B: GMMIR Profile - Horizontal Distance (m) at the Woodbine site.

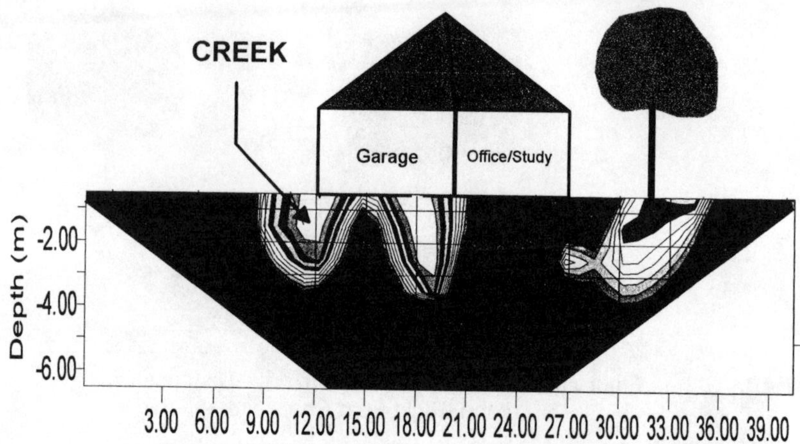

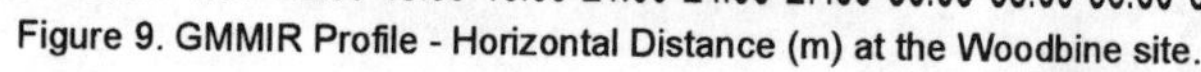

Figure 9. GMMIR Profile - Horizontal Distance (m) at the Woodbine site.

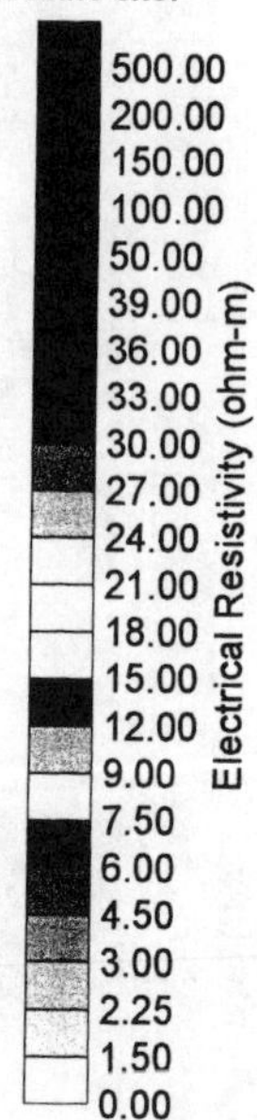

Notes: 1. Data at lower corners is interpolated.
2. Structure, Boring and Vegetation positions are approximate.
3. Patent Pending Process, All Rights Reserved.
US Patent Application S/N 09/071,577.

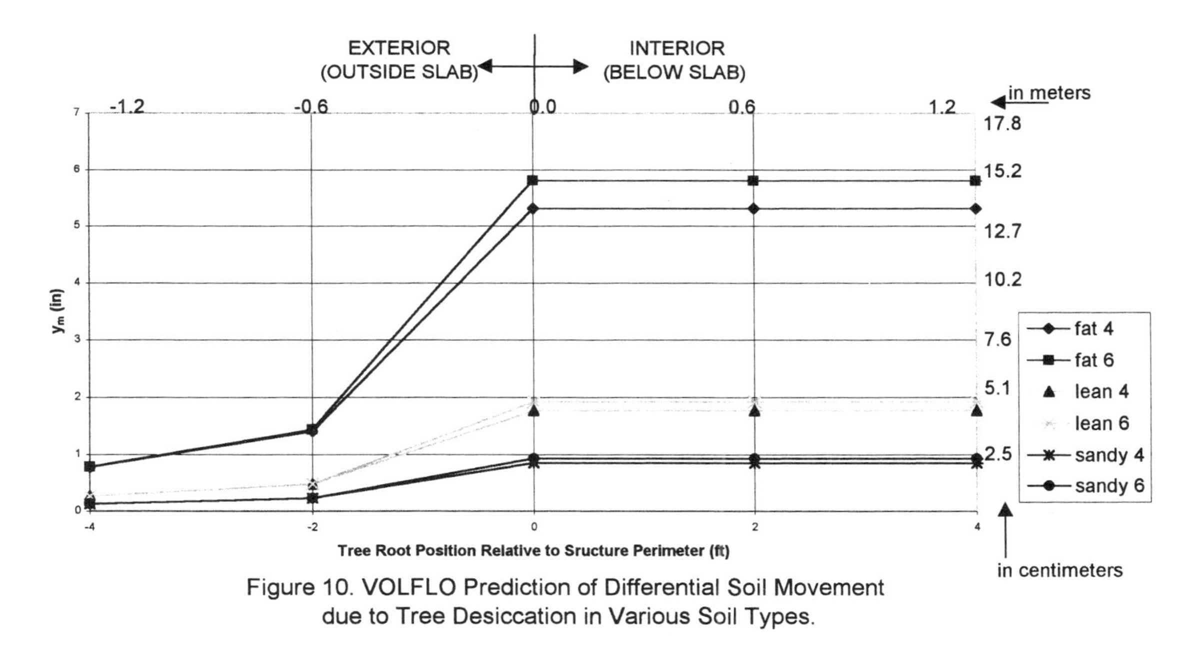

Figure 10. VOLFLO Prediction of Differential Soil Movement due to Tree Desiccation in Various Soil Types.

Longitudinal Cracking of a Bicycle Trail Due to Drying Shrinkage

James B. Nevels, Jr., M. ASCE

Abstract

The effect of unsaturated soils on pavement performance is examined through a case history involving a narrow pavement constructed on an at-grade section. The pavement function was for bicycle and pedestrian traffic in a recreational project in Oklahoma City, Oklahoma. The pavement section consisted of asphalt concrete surface course underlain by aggregate base on a prepared subgrade. Following construction, and during a very dry weather cycle, the pavement experienced a total of 3237 m of predominately longitudinal cracking and surface undulation.

A reconnaissance of the pavement extent (5.58 km) indicated that 68 percent of the cracking occurred in one pedological soil series. A geotechnical study was performed that included: pavement core and base course tests, hand auger and continuous Standard Penetration Test borings, index and engineering property tests and filter paper method soil suction testing. Examination of an average yearly water balance prior to and following pavement construction revealed that shrinkage rather than heave was the crack controlling mechanism. Estimates of field capacity were made from water balance calculations. The Thornthwaite Moisture Index (TMI) and the Soil Moisture Deficits (SMD) were estimated for the period prior to and following construction by comparisons of monthly rainfall and monthly potential evaporation resulting in indications of potential shrinkage and cracking. A characteristic curve was developed from the total suction measurements using the filter paper method. Analysis of available information indicates that the change in the total suction was responsible for soil shrinkage. Suction changes were determined through suction versus depth profiles over time to predict the initial and final suction input and using initial calculated suction from the characteristic curve(s).
Finally it was shown that the change in suction, resulting in lateral shrinkage of the soil, caused lateral forces that easily exceeded the tensile strength of the asphalt concrete and aggregate base courses.

James B. Nevels, Jr., Ph.D., Soils and Foundations Engineer, Materials Division, Oklahoma Department of Transportation, 200 N.E. 21st Street, Oklahoma City, Oklahoma 73105, Tel. (405)522-4998, email:jnevels@odot.org

Introduction

The purpose of this report was to investigate the causes of the pavement cracking and surface deformation that has occurred at this project since the end of pavement construction. The pavement function is for pedestrian and bicycle traffic. The project was authorized by the City of Oklahoma City as a recreational project that extends around the southern perimeters of Lake Hefner and through parts of the Lake Hefner golf course.

The project is located in northwest Oklahoma City. Construction plans were prepared by FHC Inc., design consultant for Oklahoma City. The supervision of the construction was handled by the Oklahoma Department of Transportation (ODOT) through Mr. Charles Thurman, then resident engineer at the Edmond office. The investigation covers the plan length of the trail pavement exclusive of the bridge length for a total length of 5.93 km. During and following construction, the asphalt pavement experienced extensive longitudinal cracking as well as differential surface deformation.

The City of Oklahoma City alleges that the ODOT contractor for this project was negligent in the construction thereby causing the cracks to develop, and that ODOT is responsible for the repair of the pavement cracking. There was no reported geotechnical study made by FHC Inc. of the pavement subgrade soil and in situ soil underlay nor was there any consideration given to the environmental factors that effect the pavement design that ODOT is aware of.

The report includes a detailed site characterization, test borings and pavement cores, laboratory soil testing, soil and pavement analysis and discussion. From an investigation management standpoint, this study concentrates on the pavement distress observed within the largest mapped soil series according to the Natural Resources Conservation Service (NRCS) Oklahoma County field office underlying the pavement namely the Renthin Soil Series and the distress related to grassroot grade level (Soil Survey of Oklahoma County, Oklahoma, 1967).

Site Characterization

The pavement section in this project is the same throughout and consists of a thin asphalt concrete surface course underlain by an aggregate base course on a prepared subgrade graded to the template shown in Figure 1. The construction sequence as related by the resident engineer consisted of the following steps:

1. Blade the top of the subgrade with a motor grader blade set at a level to cut a 2 percent cross slope. The cross slope direction alternated right or left depending upon the natural ground line slope in order to maintain drainage.
2. Place and compact the aggregate base to the planned thickness on the

constructed cross slope and level the top surface.

3. Lay the asphalt concrete surface course to a level grade.

During the rolling process, the contractor invariably, because of the type of steel-wheeled roller used, put a slight crown in the asphalt concrete surface near the pavement edges. It was estimated by the resident engineer that a 12.70 mm crown was formed on each pavement edge relative to the pavement surface. The pavement grade is predominately at a grassroot level on natural ground with slopes that vary between 0 and 3 percent. Adjacent to the pavement edges bermuda slab sod was placed. The pavement was contoured around and through some stands of existing tress. A total of 266 new trees were planted along the alignment.

The pavement distress throughout the project extent consists predominately of longitudinal cracking, see Figure 2. Using a wheeled counter and measuring in meters the total pavement length and a crack survey were made. The actual pavement length (5.93 km) checked very close to the planned pavement length (5.95 km). Following the Strategic Highway Research Program Distress Identification Manual (SHRP-P-338,1993), the longitudinal cracking can be characterized as having the following percentages of low, moderate and high severity levels respectively: 17.5, 75.6 and 6.9. A total of 3.24 km lineal feet of cracking was recorded.

The deflected shape of the pavement surface, confirmed by conventional survey rod shot elevations, measured to an accuracy of 3.05 mm at the edges, center and at one-eighth points across the pavement surface(a total of seven elevations). Six 30.478 m length test sections were surveyed in this manner within the Renthin soil series station extents to depict the differential pavement surface deformations.

In an effort to study the effect of climate on differential movements the Thornthwaite Moisture Index (TMI) (C. W. Thornthwaite , 1948) was calculated for the years prior to and following pavement construction. The TMI requires only the total monthly precipitation, average monthly temperature and north latitude of the location in order to calculate the index. The index relates the precipitation and potential evapotranspiration at monthly intervals in determining the moisture balance on an annual basis. The mean TMI for a 44-year period of record for the Will Rogers Airport weather station in Oklahoma City is 6.3. The calculated TMI for eight years preceding pavement construction is a follows: 19.5, 24.9, 17.9, 22.0, -2.2, 16.4, 10.6, 6.8 and 0.8 in 1998. The moisture balance constructed using the TMI data prior to and following the pavement construction is shown in Figures 3 through 6. The influence of weather upon soil drying can be presented in another manner where by monthly potential evapotranspiration (calculated according to the Thornthwaite method) is plotted against corresponding monthly rainfalls, see Figures 7 through 10. This type of presentation introduces the concept of Soil Moisture Deficit (SMD) whereby the SMD is the potential evapotranspiration minus the rainfall recorded weekly and plotted below the horizontal axis. A look at the prior SMD through to the present time (1990-2000) is seen in Figure 11.

The Renthin soil can physically be characterized as a residual soil formed in material weathered from clayey and silty shale of the Hennessey geologic unit (Permian geologic age) (Engineering Classification of Geologic Materials, Division 4, 1967). Further it is a shallow depth clay having medium to high plasticity, medium stiffness, blocky structure with a few slickensides, predominately a moist moisture state, mottled color and a high shrink-swell potential. The Renthin soil is overconsolidated due to desiccation. The soil taxonomy for the Renthin series is fine, mixed, superactive, thermic Udertic Arigiustolls.

TEST BORINGS

A total of six continuously sampled Standard Penetration Test (SPT) borings were made to the point of refusal to verify the Renthin soil and underlying Hennessey geologic stratiography and index tests at this site. These SPT borings were conducted according to ASTM D 1586 test procedure (ASTM, 1999). They confirm the presence of the Renthin soil profile with an average depth to the top of shale of 1.767 m. A representative Renthin series soil profile is shown in Figures 12.

Adjacent to the SPT borings and within 0.91 m, two hand auger borings and four continuously thin-walled tube sample borings were made for the purpose of taking samples for index, wet density and soil suction testing. A further seven auger borings were made through 10.16 cm diameter core holes in the pavement for the purpose of checking the pavement section (asphalt concrete and aggregate base) thickness, indirect tensile strength testing and subgrade soil sampling. All auger and continuous thin-walled tube borings were conducted according to ASTM D 2113 and D 1587 test procedures (ASTM, 1999) respectively.

Fourteen 10.160 cm diameter pavement cores were cut in total; seven at borings 9 through 15, and seven randomly selected through longitudianl cracks. Seven cores with no cracks were tested to determine the indirect tensile strength of the asphalt concrete according to AASHTO T 283(AASHTO, 2000). During the course of this study no perched groundwater tables were observed in the borings. The moisture state varied from moist to dry to wet.

Laboratory Soils Testings

The laboratory analysis consisted of two parts. The first part was to document the index and physical properties of the Renthin soil series and underlying soil and weathered shale. The liquid limit, plastic limit, shrinkage limit, moisture content and wet density were performed according to ASTM D 4318, D 4959 and AASHTO T233 (ASTM 1999; AASHTO 2000) respectively. The second part was to study the in situ suction and shrinkage for the same.

The site stratiography and index properties typical of these station extents are presented in Figure 12. The variation in moisture content and wet unit weight versus depth for various borings is shown in Figure 13.

The ASTM D 5298-94 test procedure (ASTM, 1999) was used to measure total suction from auger samples taken with depth. The procedure used deviated from the D 5298-94 apparatus requirements in the following four ways:

a. An equivalent Schleicher-Schnell No. 589 filter paper (VWR Grade No. 74,55mm) was used. It was not specially pretreated.
b. A 230-ml polyethylene specimen container having a screw-on lid with an o-ring seal was used instead of a metal or glass container.

c. Two wraps of making tape (12.7mm wide) were used instead of flexible plastic electrical tape to further seal the outside lid/container connection.
d. Expandable polyethylene netting was used instead of screen wire or brass discs.

All ASTM D 5298-94 procedure requirements were met. Soil samples for total suction testing were taken from thin-walled tube samples and from auger cuttings. The filter paper moisture contents from these samples were converted to suction values using both the upper and lower segments of the Scheicher and Schuell No. 589 calibration curve given in ASTM D 5298 test procedure (ASTM 1999). A characteristic curve was developed for both total and matric suction for samples representative of the B-horizon of the Renthin soil series from thin-walled tube samples taken in an offset boring from boring 1, see Figure 14. Total suction versus depth profiles were developed for various boring locations between April and July 1998 during the peak of the 1998 drought, reference Figure 15. The suction profile for B-1 in Figure 16 was tested from samples retrieved following a very wet period (in May 1999) where the soil profile for a short period was much wetter.

Shrinkage typical of the Renthin series B-horizon was developed from thin-walled tube samples taken from the same offset boring. A shrinkage curve representative of the B-horizon is given in Figure 16. The method used to develop this curve was the basic CLOD shrinkage test procedure by (McKeen, 1981) furthered explained by (Nelson and Miller, 1992).

Soils and Pavement Analysis and Discussion

It has been established through the walkout of the project that the grade is predominately at a grassroot level in reference to the natural ground, and that the ground slope conforms to the soil series map unit slope of between 0 and 3 percent.. There were only a few short station extents where the existing trees obviously influenced the cracking. Approximately 75 percent of the 266 newly planted trees were within the Renthin soil extents. All were discounted as having any influence on the cracking due to their size, location in most cases, being placed in a burlap ball, the tightness of the soil and the length of time being planted in the ground.

The predominate defect is the longitudinal cracking and is characterized as having moderate severity based on the SHRP-P-338 criteria. The positioning of these longitudinal cracks was generally along the pavement edge. An analysis of the Renthin soil series, cross-checked with the crack survey, reveals that 68 percent of the total pavement cracking occurred within the Renthin station extents. The profiles for the six test sections appear to indicate a predominate center lift of the pavement similar to that reported (Wray, 1987; Nelson and Miller 1992) for building slabs on grade. Results of the survey indicate a range of differential movement from 0 to 38.1 mm. There is also some indication of a combination of edge and center lift in all six test section areas. Previous research has normally supported the concept of the center lift case resulting from increased moisture content as a result of elimination of evapotranspiration and represents a severe long-term distortion condition. However, it is our belief that the same effect can occur due to skrinkage or dessication along the pavement edge especially in the case of such narrow width pavement.

The calculated yearly TMI data for the eight years prior to construction shows that there is a general overall drying trend in reference to the mean annual historical TMI. In particular a plot of the moisture balance for the years 1997 to 2000 (Russam and Coleman, 1961) reveals the more long term cyclic effect of precipitation and evapotranspiration, see Figures 3 through 6. In reference to this project as can be seen in Figure 4, the construction time line shows that the pavement was constructed in a wet season followed by a significant drying out period. This period locally was termed the 'Drought of 1998' in central Oklahoma. Presented in Figure 7 through 10 is the new concept by (Biddle, 1998) where rainfall and potential evapotranspiration are plotted on the vertical axis against time on the horizontal axis. The moisture deficit is the difference between the two.Our analysis differs from Biddles in that the potential rather than the actual evapotranspiration is used, the Thornthwaite evapotranspiration is used instead of Penman's and a monthly basis of calculation is used rather than a weekly. For this narrow width pavement, the moisture content in the center of the pavement does not appear to be greater than along the outside edge and further out lying areas as can be seen in depth for various site locations in the Renthin series, see Figures 11 and 12. The edge moisture variation distance 'E' as defined by Wray, 1980 is the distance measured inward from the slab edge (pavement edge in this case) over which soil moisture varies enough to cause pavement movement. According to Wray, 1980; McKeen and Johnson, 1990 typically 'E' ranges from 0.61 m to 1.83 m and can equal the active zone (defined later).

Borings confirm that the residual soil profile of the Renthin soil is relatively shallow and with a consistent depth to bedrock, see Figure 12. The active zone (depth for which moisture content and soil suction can vary) is considered the full depth of the Renthin soil series profile and further down to the top of the weathered Hennessey shale, refer to Figure 12.

The taxonomy classification for the Renthin soil indicates the following significant facts:

a. The term 'fine' implies that it is clayey.

b. The term 'mixed' refers to a mineralogy classification that means no one clay mineral predominating. One can expect a mix of expansive clay minerals in the soil solum.

c. The term 'superactive' implies a high cation exchange capacity and again an expansive clay mineralogy.
d. Udertic implies vertic properties for a mollisol. Generally cracks form within 1.245 m from the top of the ground and are 0.508 cm or more wide through a thickness of 30.48 cm or more for some time in most years. Unless irrigated these cracks remain open for at least 90 cumulative days per year.

The 'ss' designation in the Renthin series noted in boring 1 profile indicates the presence of slickensides in the 'B' subhorizons which is another indicator of high shrink-swell potential.

The central question here is what was the initial total suction profile versus depth prior to construction. We can deduce from the moisture balance for the fall of 1997 and early winter of 1998 shown in Figures 3 and 4 that the construction was initiated in a wet season. It appears that there is little difference between the total suction profiles in Figure 15 and 16 although the suction band is slightly lower for the profiles underlying the pavement as opposed to these outside of the pavement area. However the B-1 profile shown in Figure 15 represents one that was developed following a very wet season. Near the soil surface, the total suction approximates a range from 100 to 1500 kPa. A review of Figures 3 through 11 seem to indicate that seldom does the soil profile reach its field capacity which by definition is a situation where by the soil profile above a watertable is allowed to drain naturally, a condition that is equivalent to a suction of 5 kPa. The answer to the question above then is indeterminate. Suctions are greatly reduced following wet seasons and it appears that the B-1 suction profile found during the course of this study is close to the probable initial total suction profile.

It is known that when horizontal shrinkage occurs there is also a reduction in the horizontal stresses within the soil mass; once these approach zero a vertical crack will develop along any pre-existing planes of weakness (Biddle, 1998). An approximate expression for the suction that will cause cracking in terms of a given vertical total stress is given by Ridley and Patel (1998):

$$\sigma_n = K_o \sigma_v + u(1\text{-}K_o) \qquad (1)$$

where σ_n , σ_v = horizontal and vertical stress

K_o = coefficient of earth pressure at rest

u = total suction

when σ_n reduces to zero the suction in equation (1) is given by

$$-u = \sigma_v \frac{K_o}{1-K_o} \qquad (2)$$

Using a K_o derived from previous dilatometer test data for these residual soils, the suction required to initiate a shrinkage crack is in the order of 650 kPa. The charge in total suction near the surface of the Renthin soil is approximately from 100 to 1500 kPa based on the open field total suction profile in Figure 15. The average tensile strength for the asphalt core is 551 kPa. Thus we may conclude that the lateral stress generated by shrinkage of the Renthin soil at or very near the surface due to an approximate total suction change of 400 kPa is more than enough to initiate soil cracking and exceed the tensile strength of the asphalt concrete surface. For this change in suction o f 1400 kPa the volumetric moisture is reduced by 8.4 percent refer to Figure 14. Using this volumetric moisture change the void ratio for the soil mass is reduced from 0.71 to 0.55 which helps to explain some of the distortion in the pavement surface.

Conclusion

What can be derived from this study are two key points:

1. It has been established that the crack causing mechanism in the pavement is the clay soil shrinkage based on the soil taxonomy, moisture balance, soil moisture deficit, and the deflected shape from the survey data. We know from studying the pedological soil survey descriptions that the Renthin series, because of its udertic sub group classification, is known to have subsurface-initiated reversible cracks form when an appreciable reduction in water content from its 'field capacity' or close to it occurs.
   The soil moisture balance and soil moisture deficit indicate the clear and cyclic effect of the climate. For the year 1998 a tremendous drought was experienced in late spring, summer and early fall. We suspect that when the pavement construction was started the Renthin soil had not reached its field capacity and before project completion the effect of the soil moisture deficit was already starting to affect pre-existing closed cracks.
2. The pavement cracking and surface distortion was due to 'free field shrinkage' due primarily to change in the total suction. As a point of reference the full field capacity of a soil can be defined as equivalent to a suction of 5 kPa whereas the maximum suction which can be experienced by the roots of plants and trees at their wilting point is about 1500 kPa. From the physical evidence of the 1998 drought, the grass did turn yellow and die and the trees did drop their leaves to a considerable degree, indicating a wilting point.

Acknowledgment

The author wishes to extend a special thanks to Kathy Kirkpatrick for typing this manuscript and to Christopher Clarke for preparing the figures.

References

AASHTO (2000), Twentieth Edition, Part II Tests.

ASTM (1999), Section 4, Volume 04.08 and 04.09, Soil and Rock (I) (II).

Biddle, P.G. (1998). Tree Root Damage to Buildings, Vol. 1, Willowmead Publishing Ltd., Ickleton Rd, Wantage, OX129JA.

Engineering Classification of Geologic Materials, Division 4, Oklahoma Highway Department, Research and Development Division, 1967.

Nelson, J.D. and Miller, D.J. (1992). Expansive Soils, John Wiley & Sons, Inc., New York.

McKeen, R. G. and Johnson, L.D. (1990). Climate-Controlled Soil Design Parameters for Mat Foundations, J. Geotech. Eng., ASCE 116(7): pp. 1073-1094, July.

McKeen R. G. (1981). Design of Airport Pavements for Expansive Clays, U. S. Dept. of Transportation, Federal Aviation Administration, Rep. No. DOT/FAA/RD-81/25.

Ridley, A. M. and Patel A. R. (1998), Tensiometers; Their Design and Use for Civil Engineering Purpose, Geotechnical Site Characterization, Volume 2, pp. 851-856.

Russam, K. and Coleman, J. D. (1961). The Effect of Climatic Factors on Subgrade Moisture Conditions, Geotechnique Vol. 11, No.1, pp. 22-28.

Soil Survey of Oklahoma County, Oklahoma (1967), USDA, Soil Conservation Service.

SHRP-P-338 (1993), Distress Identification Manual for the Long-Term Pavement Performance Project, Strategic Highway Research Program, National Research Council, Washington D. C.

Thornthwaite, C. W. (1948). An Approach Toward a Rational Classification of Climate, Geographical Review, Vol. 38, pp. 55-94.

Wray, W. K. (19987). The Effect of Climate on Expansive Soils Supporting On-

Grade Structures in a Dry Climate, Transportation Research Record No. 1137, TRB, National Research Council, Washington D. C. pp. 12-33.

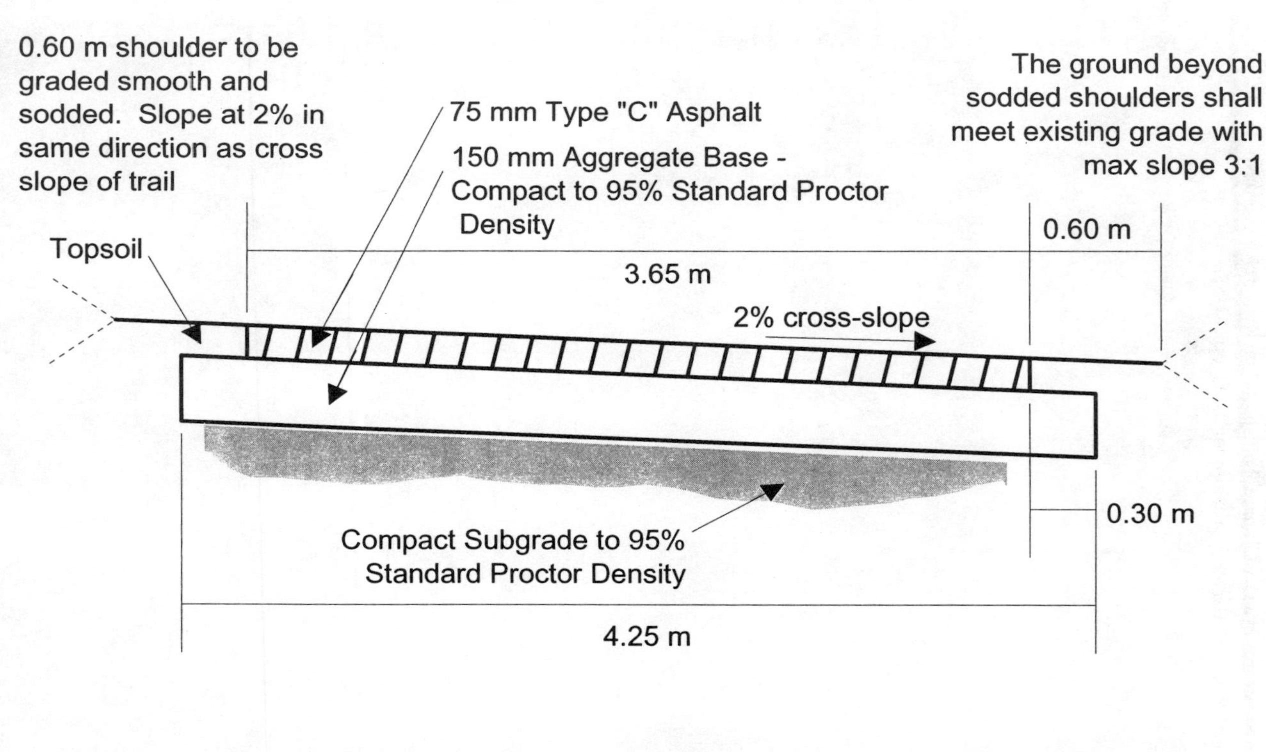

Figure 1. Typical Bicycle Trail Pavement Section

Figure 2. Example of Longitudinal Cracking at Station 0+348 (metric)

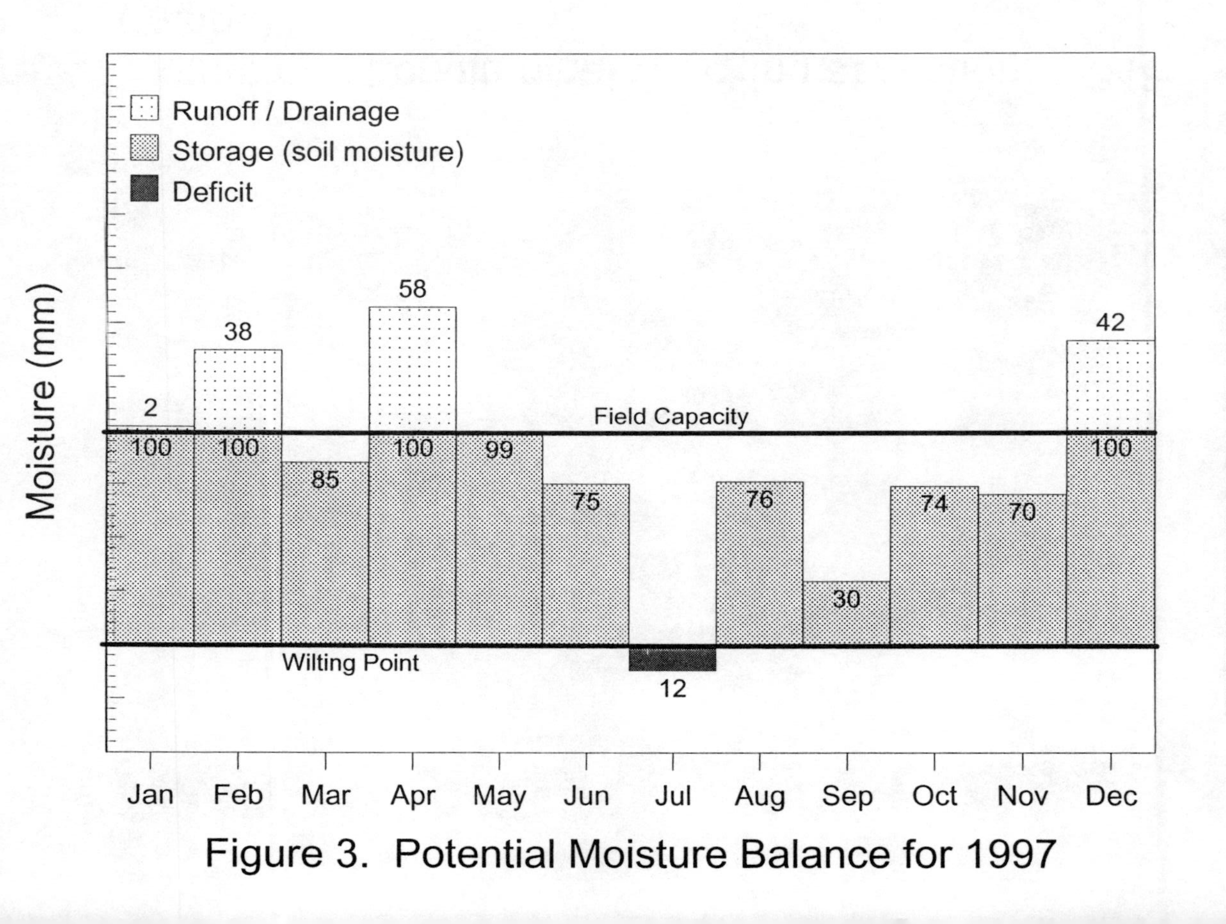

Figure 3. Potential Moisture Balance for 1997

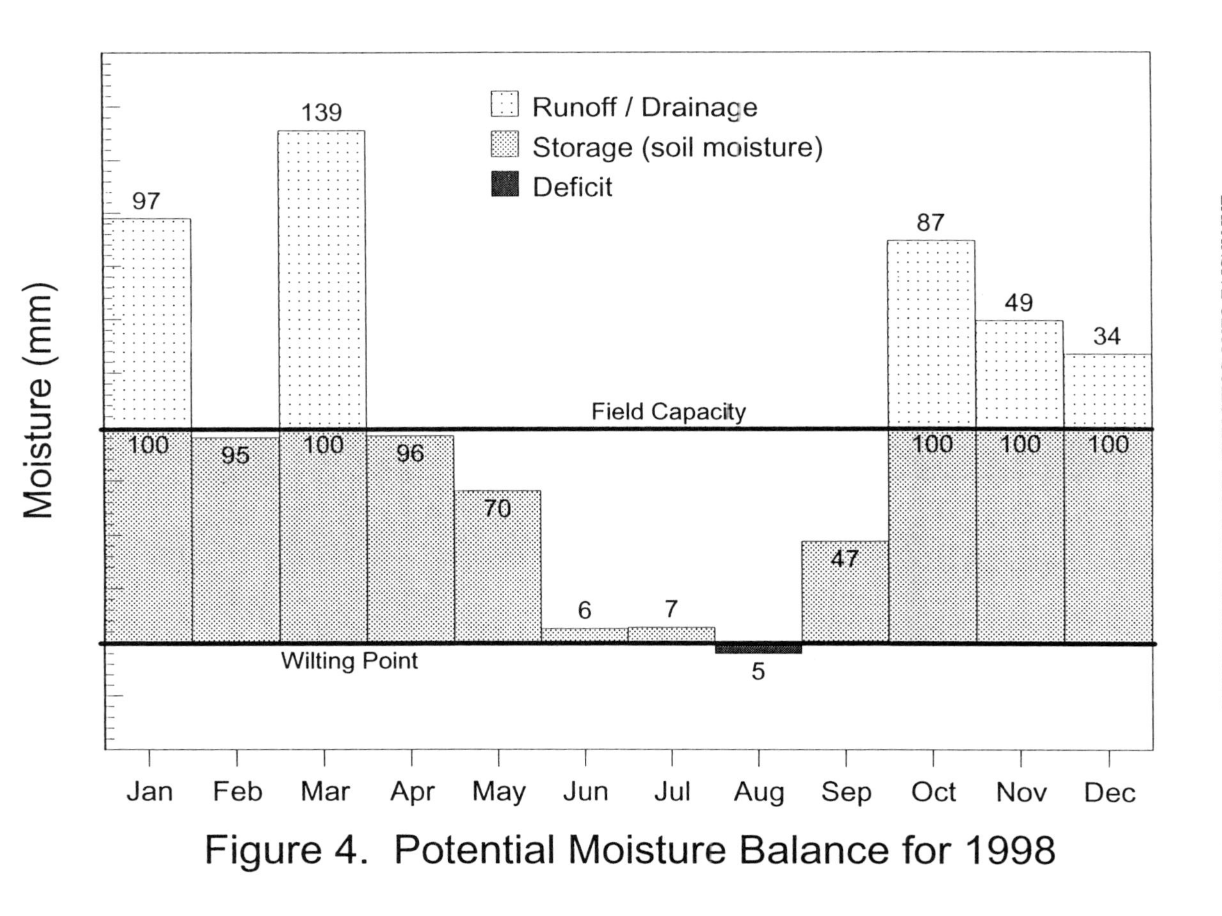

Figure 4. Potential Moisture Balance for 1998

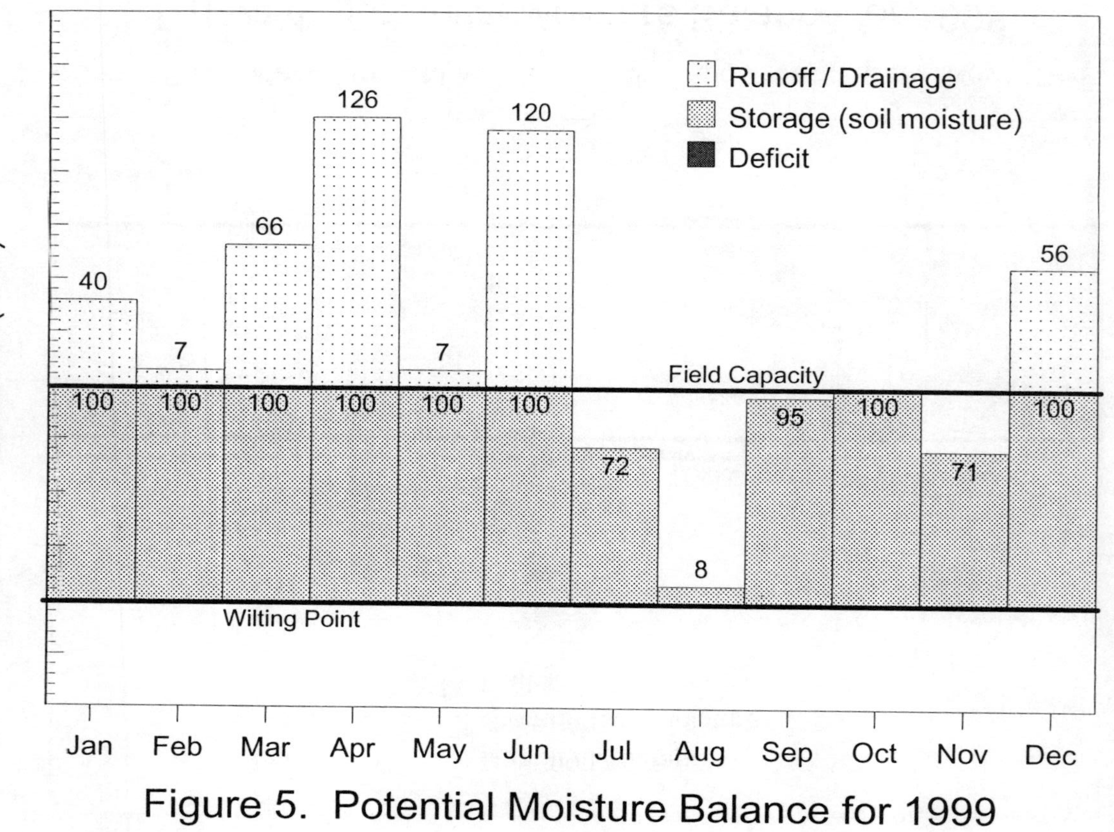

Figure 5. Potential Moisture Balance for 1999

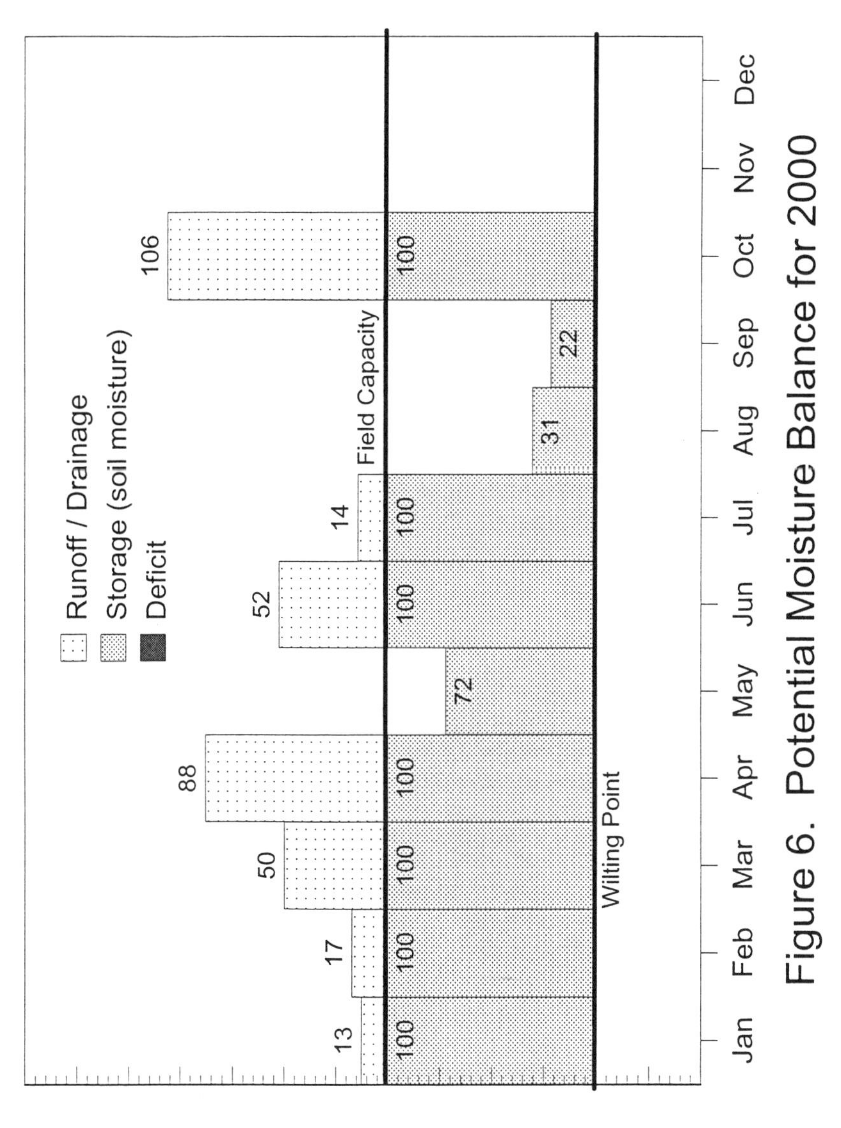

Figure 6. Potential Moisture Balance for 2000

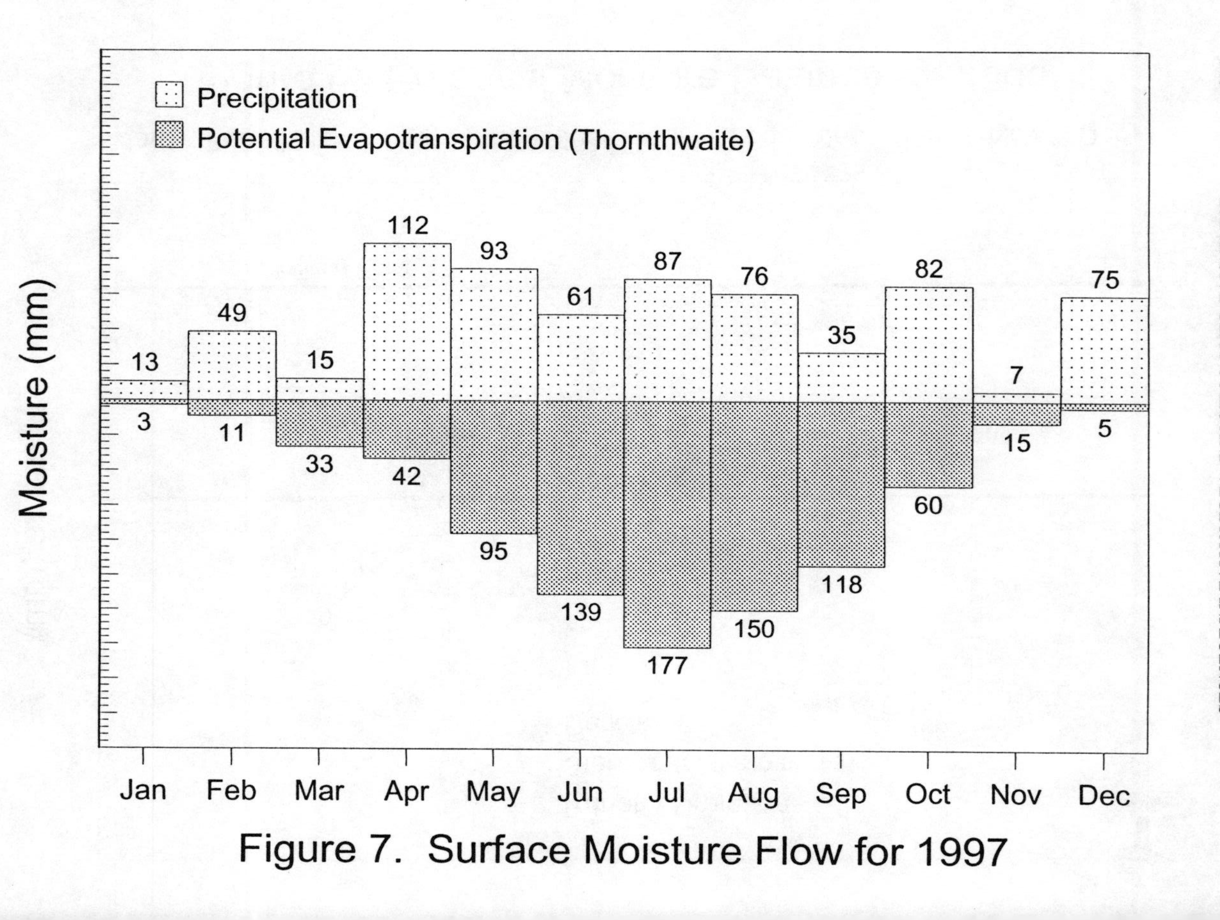

Figure 7. Surface Moisture Flow for 1997

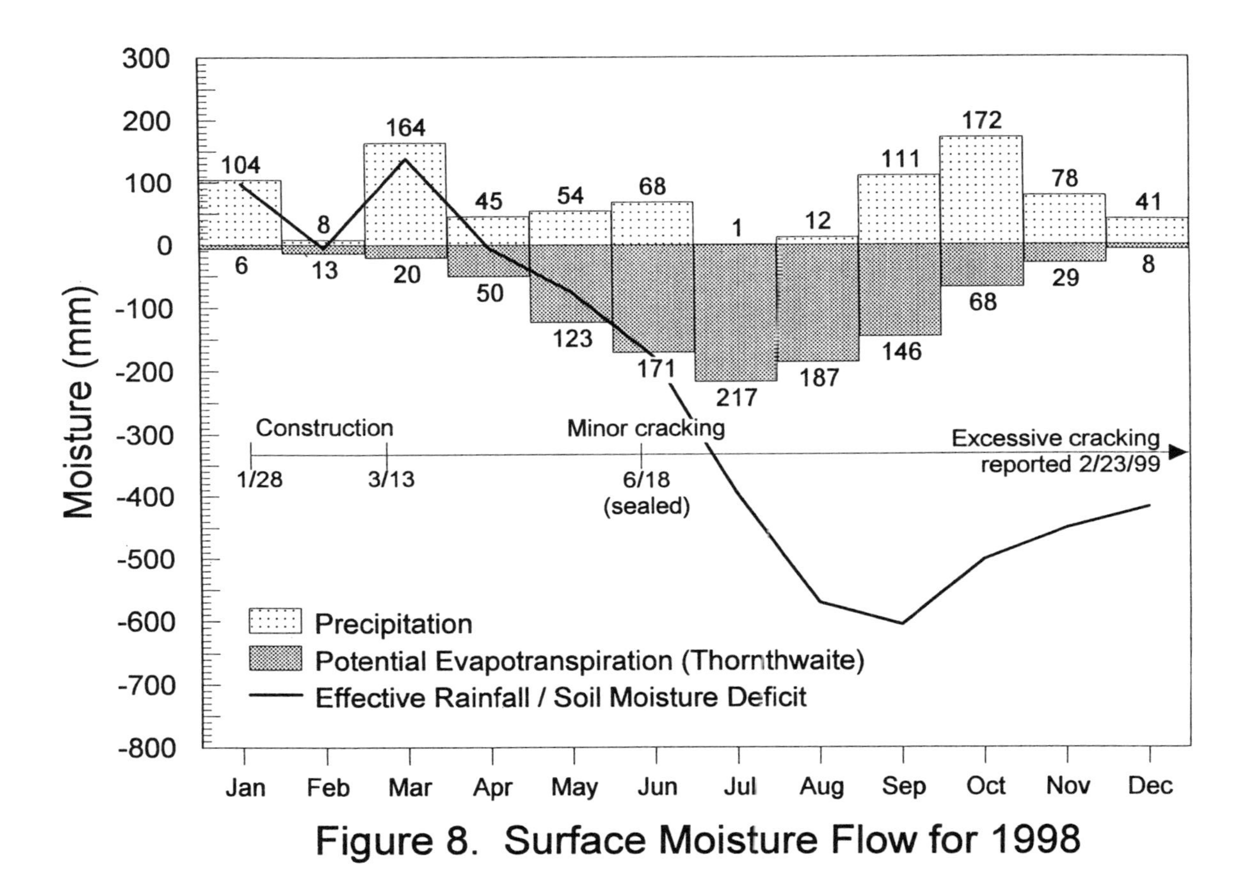

Figure 8. Surface Moisture Flow for 1998

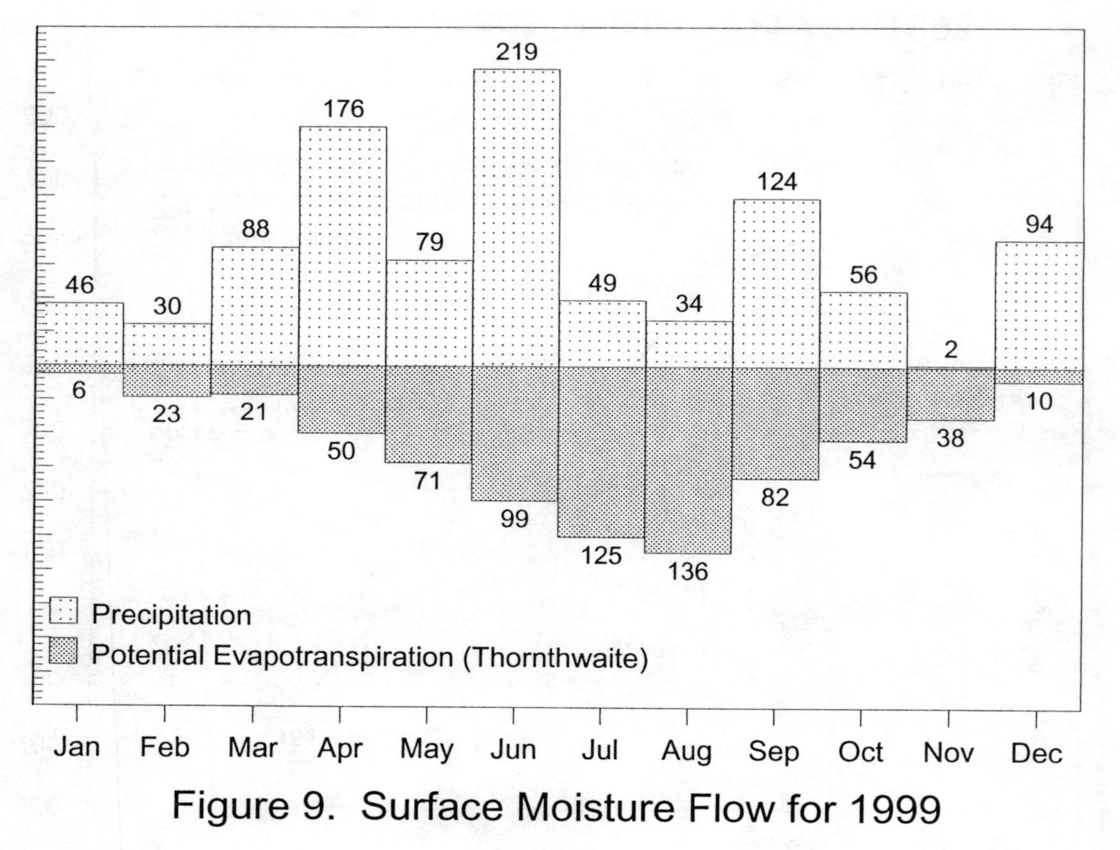

Figure 9. Surface Moisture Flow for 1999

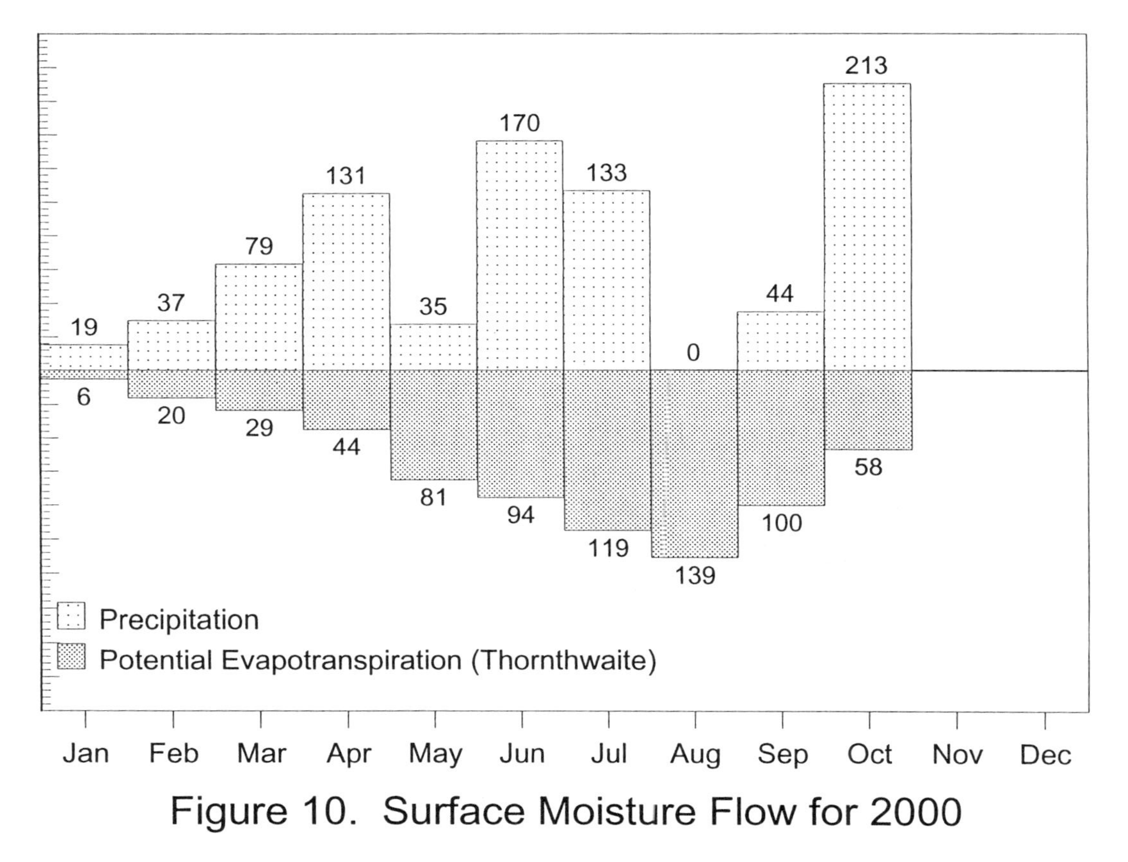

Figure 10. Surface Moisture Flow for 2000

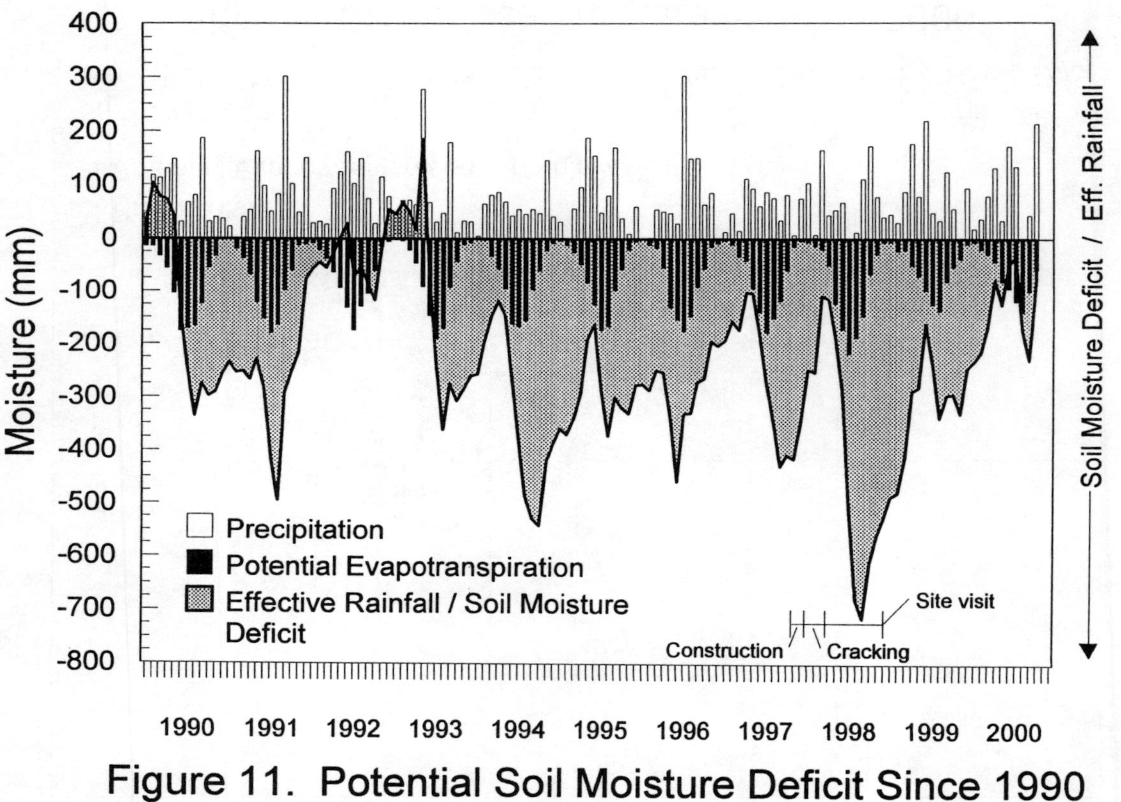

Figure 11. Potential Soil Moisture Deficit Since 1990

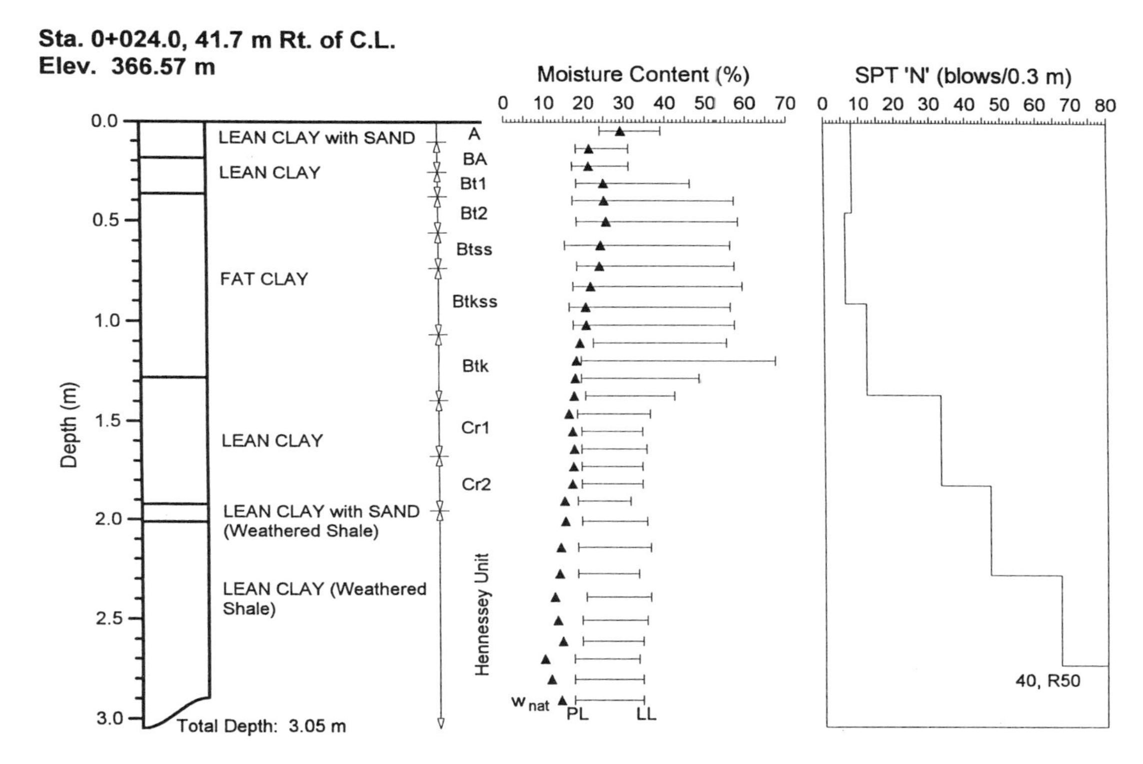

Figure 12. Boring No. 1, Typical Renthin Series Profile

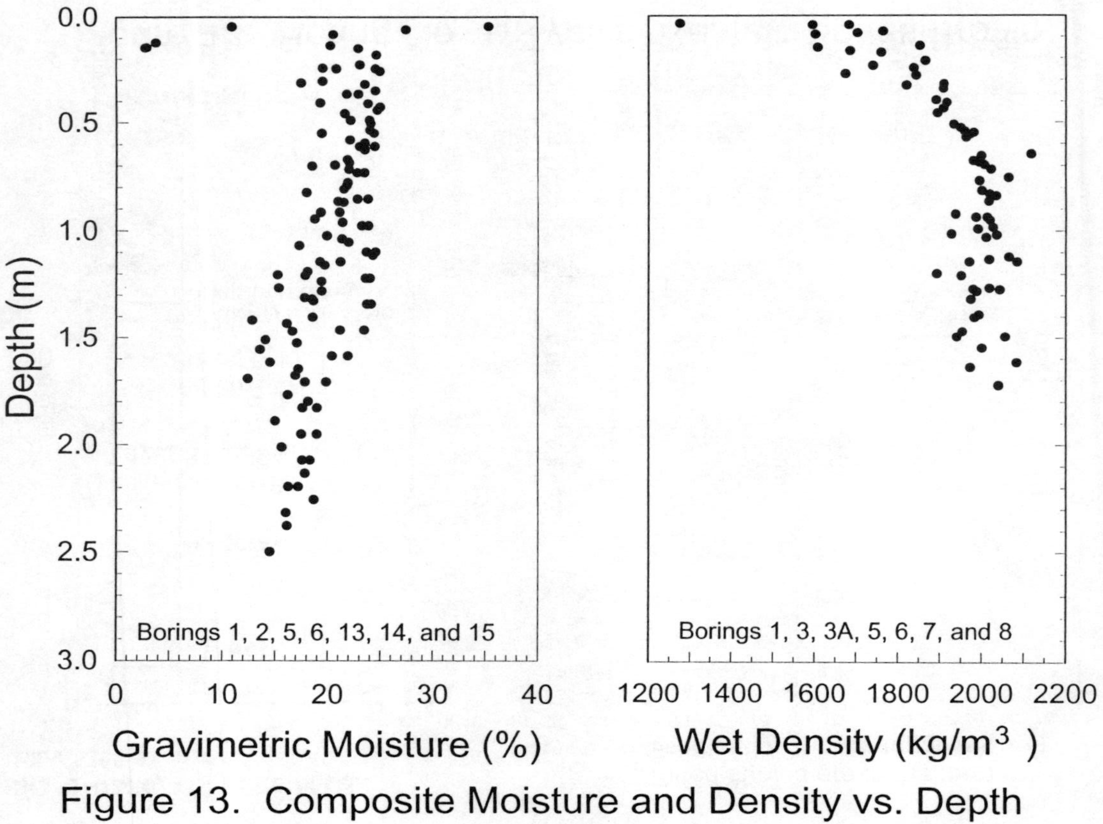

Figure 13. Composite Moisture and Density vs. Depth

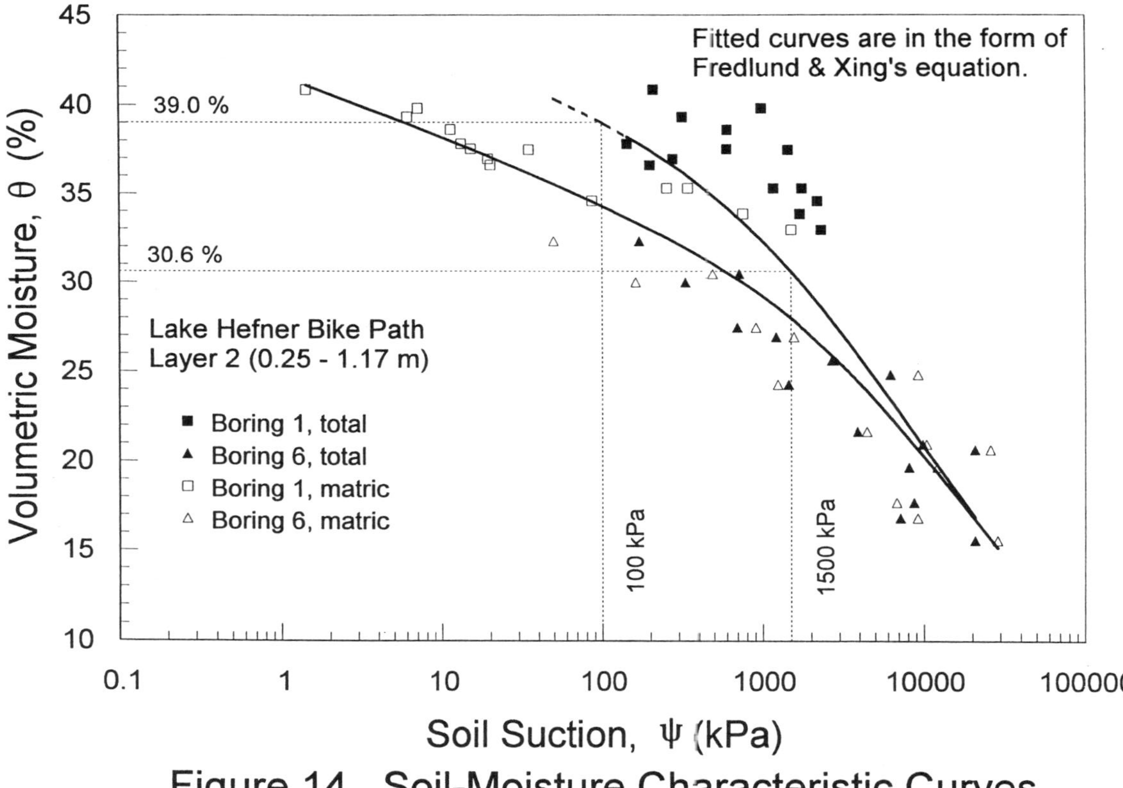

Figure 14. Soil-Moisture Characteristic Curves

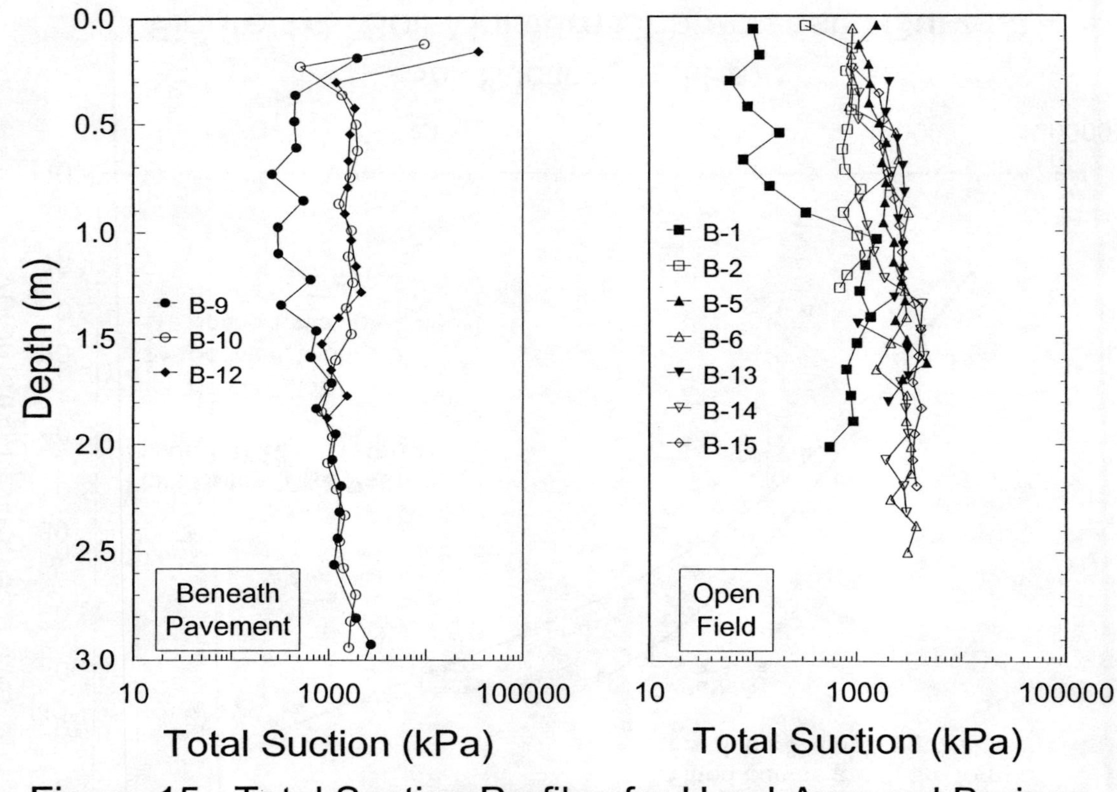

Figure 15. Total Suction Profiles for Hand-Augered Borings

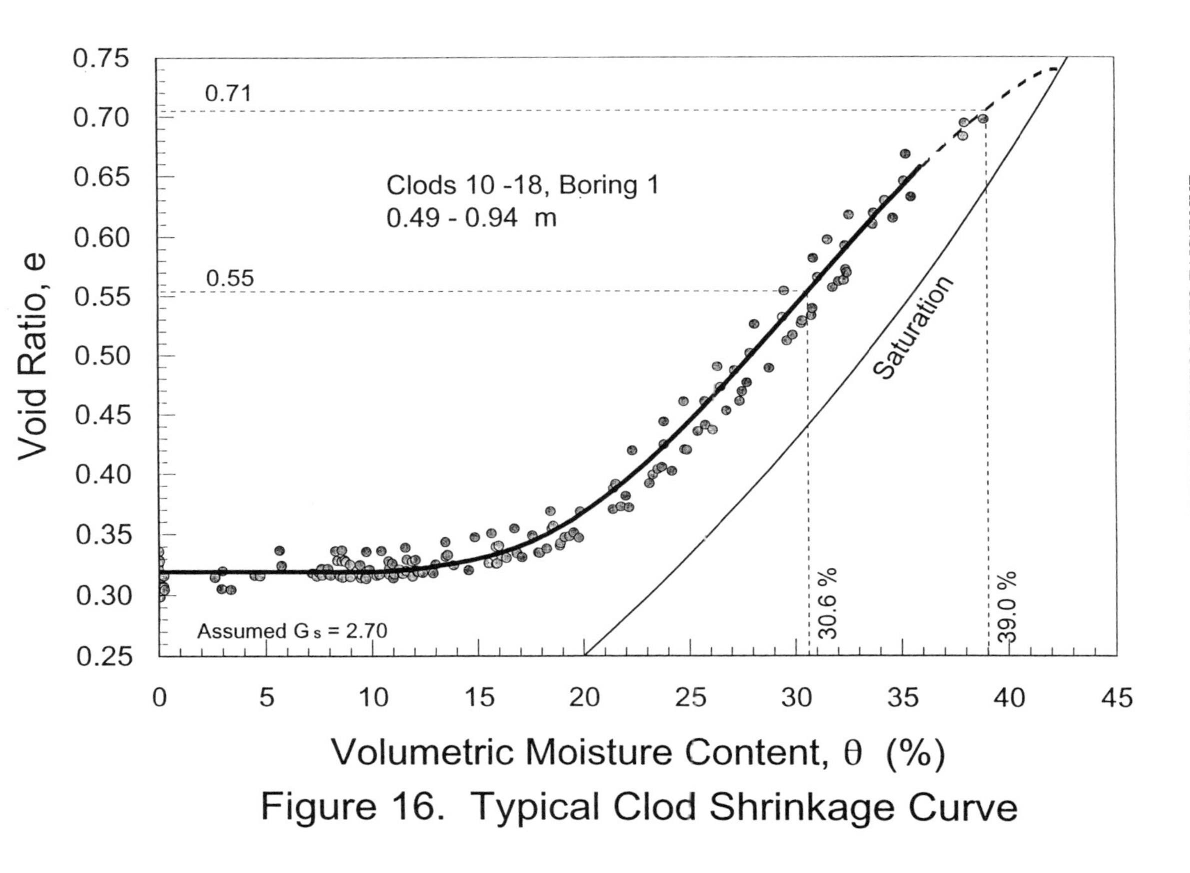

Figure 16. Typical Clod Shrinkage Curve

Influence of Local Tree Species on Shrink/Swell Behavior of Permian Clays in Central Oklahoma

Donald R. Snethen, PhD, PE[1]
Fellow, ASCE

Abstract

The central region of Oklahoma is underlain by sedimentary deposits of the Permian period, which include shales and related materials. The fine-grained materials weathered from these Permian shales are typically low to moderate plasticity clayey silts, silty clays, and clays. These fine-grained materials exhibit low to moderate shrink-swell potential with some local areas showing high shrink-swell potential. The general climate of the region is dry subhumid to semiarid with climatic extremes, as was the case during the summer of 2000, which was one of the driest on record. The influence of climate on the behavior of moderate to high shrink-swell potential soils can be more extreme, when the influence of vegetation is added. This paper looks at the occurrence of damage caused by shrinking and swelling soils and the role that vegetation had in causing the problem to be more severe. Some basic physiological properties of common species of trees in the region as well as local characteristics of the soils are reviewed. Case studies of shrinking and swelling soil damage to streets and residential construction are presented and discussed.

Background

The shrink-swell characteristics of Permian soils in central Oklahoma have been documented by various authors (Haliburton and Marks, 1969; Snethen, 1984, 1992). Permian soils in this region are typically fine-grained residual materials weathered from claystone or siltstone parent rock. The resulting soils are clayey silts, silty clays, and clays with low to moderate plasticity characteristics. The magnitude of shrink/swell effects of the Oklahoma Permian soils pale in comparison to such materials as the Yazoo clay of Mississippi, Taylor marl of Texas, and Pierre shales of Colorado and Wyoming; however, measured deformations of pavements and lightly loaded structures have exceeded 50 mm.

---

[1]Professor, School of Civil and Environmental Engineering, ES 207, Oklahoma State University, Stillwater, OK 74078-5033.

Shrink/swell potential of swelling soils is determined by the clay mineral characteristics of the soil (i.e., type and amount) and is typically reflected in such measured intrinsic properties as soil suction or plasticity. The extent to which shrink/swell potential of swelling soils is realized is dependent on such extrinsic factors as initial moisture content and dry density, availability of moisture, climate conditions and vegetation (type and proximity to structure). Initial moisture content and dry density conditions are factors that can be characterized early in the design/construction process, but are often not appropriately measured or accounted for in a design. For example, subsurface exploration and laboratory testing programs are often conducted when moisture conditions are high, as in late Winter and Spring, but actual construction often begins at times when moisture is depleted, late Summer and Fall. This sequence then "builds in" a depleted moisture condition beneath a structure, to which the soil responds by accumulating moisture and swelling. Obviously, a better understanding of the influence of climate on the moisture regime of the supporting soil would help minimize the amount of swell and associated damage.

The availability of moisture during and following construction is likewise influenced by climate. Figure 1 shows the mean annual Thornthwaite Moisture Index (TMI) for the state of Oklahoma based on observation records from 1944-1990. The TMI is a climate classification, which incorporates precipitation and potential evapotranspiration into the system. Generally accepted standards for TMI climate classification are:

$20 \le TMI \le 100$	Humid
$0 \le TMI \le 20$	Moist Sub-Humid
$-20 \le TMI \le 0$	Dry Sub-Humid
$-40 \le TMI \le -20$	Semi-Arid
$TMI \le -40$	Arid

Oklahoma has a wide range of climates from Humid to Semi-Arid. In Central Oklahoma, the climate classification varies between Moist Sub-Humid and Dry Sub-Humid.

In Central Oklahoma, subgrade moisture accumulates during late Winter and Spring (typically January through June) and depletes during Summer and Fall (typically July through December). Normal climatic conditions (monthly average temperature and monthly total precipitation) for the entire state of Oklahoma for 1999 and 2000 are shown in Figure 2 (Oklahoma Climatological Survey, 2000). Records show that, statewide, average monthly temperatures were at or above normal for July, August, and September for both years; and total precipitation was at or below normal for the same period in both years. August and September, 2000, were record conditions for both temperature and precipitation, with the precipitation for August being the lowest in the 109 years of records. These variations from normal conditions are significant enough to mobilize shrink/swell potential of local soils and result in damage to structures that had previously been unaffected by volume change. Specific climate observation records for the case study locations are presented in Tables 1 and 2 (Oklahoma Climatological Survey, 2000).

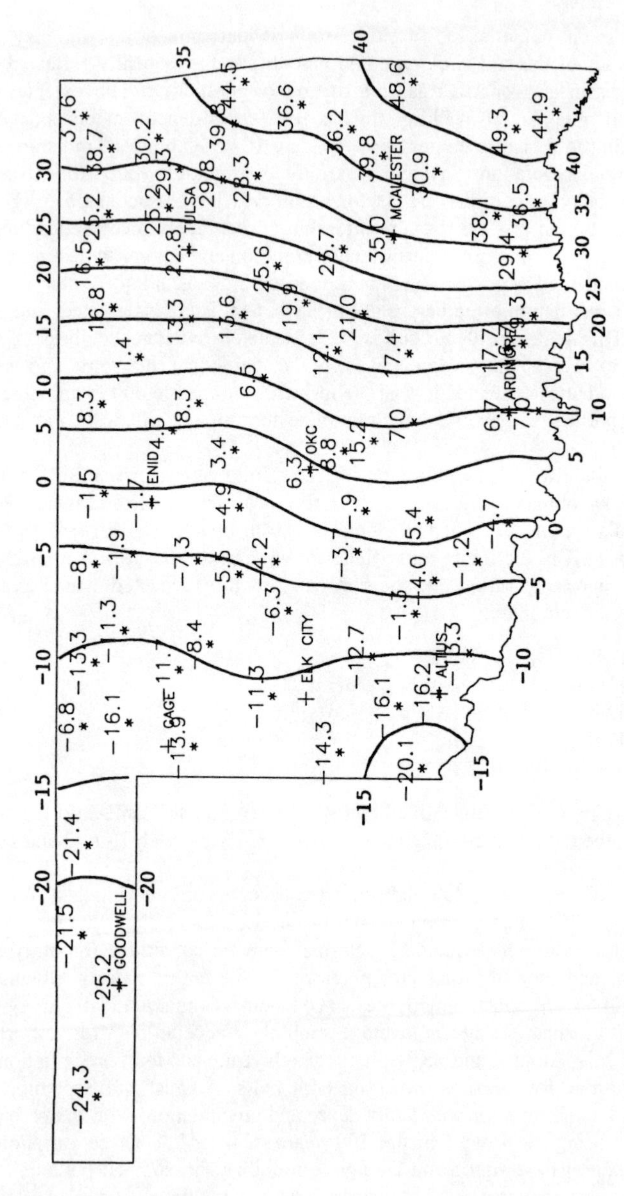

Figure 1. Mean Annual Thornthwaite Moisture Index for Oklahoma (1944-1990 Records) (Oklahoma Climatological Survey) and Dr. Jim Nevels, Oklahoma Department of Transportation

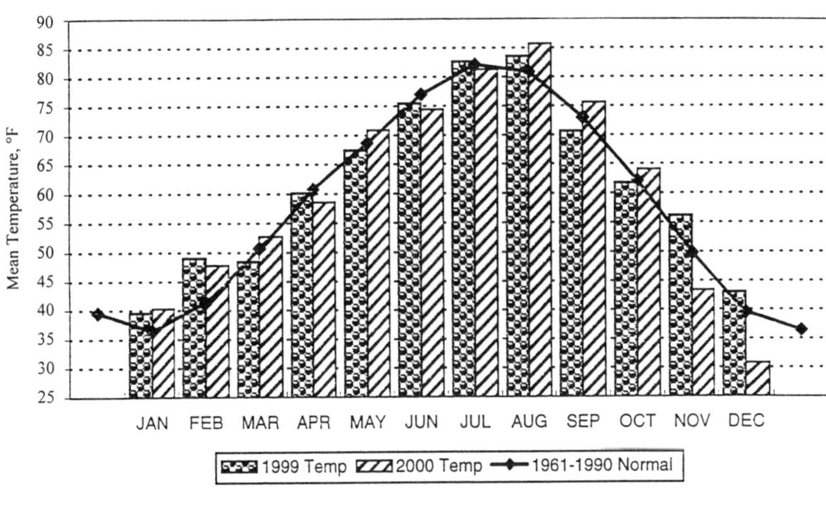

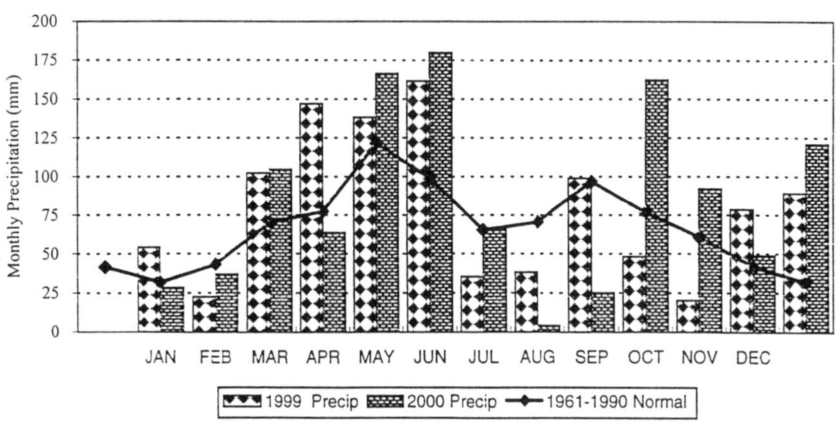

Figure 2. 1999 and 2000 Statewide Temperature (Upper) and Precipitation (Lower) Values for Oklahoma (Oklahoma Climatological Survey)

Table 1. Mean Monthly Temperature and Total Monthly Precipitation for Stillwater, OK—1999 and 2000 Observations

Month	Mean Temp (°F)	Normal Temp (°F)	Deviation From Normal	Total Monthly Precipitation (mm)	Normal Precipitation (mm)	Deviation From Normal
2000						
January	39.4	33.6	+5.8	19.8	29.2	-9.4
February	46.0	39.0	+7.0	28.5	38.9	-10.4
March	52.4	48.2	+4.2	101.9	70.9	+31.0
April	59.3	59.3	0	46.8	74.2	-27.4
May	70.3	67.3	+2.6	166.9	130.3	+36.6
June	74.2	76.2	-2.0	193.1	101.7	+91.4
July	80.9	81.6	-0.7	130.6	73.7	+56.9
August	85.8	80.3	+5.5	1.3	70.1	-68.8
September	75.5	72.1	+3.4	1.3	109.0	-107.7
October	64.6	60.6	+4.0	156.0	71.9	+84.1
November	44.4	48.5	-4.1	52.1	57.2	-5.1
December	29.7	37.9	-7.7	39.4	33.1	+6.3
1999						
January	40.3	33.6	+6.7	40.2	29.2	+11.0
February	46.8	39.0	+7.8	16.3	38.9	-22.6
March	47.3	48.2	-0.9	116.4	70.9	+45.5
April	60.3	59.3	+1.0	171.6	74.2	+97.4
May	67.9	67.3	+0.6	114.9	130.3	-15.4
June	75.4	76.2	-0.8	206.5	101.7	+104.8
July	82.6	81.6	+1.0	49.1	73.7	-24.6
August	84.1	80.3	+3.8	24.6	70.1	-45.5
September	70.1	72.1	-2.0	155.5	109.0	+46.5
October	61.4	60.5	+0.8	90.9	71.9	+19.0
November	56.0	48.5	+7.5	7.6	57.2	-49.6
December	42.9	37.9	+5.0	127.8	33.1	+94.7

Normals based on 1961-1990 records.

Table 2. Mean Monthly Temperature and Total Monthly Precipitation for Kingfisher, OK—1999 and 2000 Observations

Month	Mean Temp (°F)	Normal Temp (°F)	Deviation From Normal	Total Monthly Precipitation (mm)	Normal Precipitation (mm)	Deviation From Normal
2000						
January	38.8	36.0	+2.8	15.1	25.4	-10.3
February	45.8	40.8	+5.0	29.0	35.3	-6.3
March	52.4	50.9	+1.5	97.5	57.2	+40.3
April	57.8	61.1	-3.3	54.9	66.3	-11.4
May	70.2	69.4	+0.8	146.3	117.6	+28.7
June	74.8	78.1	-3.3	263.4	108.5	154.9
July	81.9	83.4	-1.5	84.1	52.1	+32.0
August	86.6	81.9	+4.7	2.3	68.6	-66.3
September	76.4	73.9	+2.5	0.5	104.6	-104.1
October	64.7	62.7	+2.0	155.4	59.2	+96.2
November	43.5	49.6	-6.1	52.1	48.5	+3.6
December	30.0	39.1	-9.1	46.3	30.3	+16.0
1999						
January	37.5	36.0	+1.5	48.5	25.4	+23.1
February	47.7	40.8	+6.9	8.1	35.3	-27.2
March	47.1	50.9	-3.8	117.1	57.2	+59.9
April	59.3	61.1	-1.8	102.9	66.3	+36.6
May	67.9	69.9	-1.5	78.2	117.6	-39.4
June	75.4	78.1	-2.7	212.3	108.5	+103.8
July	82.5	83.4	-0.9	79.2	52.1	-27.1
August	84.5	81.9	+2.6	26.4	68.6	-42.2
September	71.3	73.9	-2.6	58.7	104.6	-45.6
October	62.6	62.7	-0.1	40.4	59.2	+18.8
November	56.4	49.6	+6.8	12.4	48.5	-36.1
December	43.0	39.1	+3.9	108.0	30.3	+77.7

Normals based on 1961-1990 records.

The interaction between vegetation and available moisture in shrinking/swelling soils can mobilize inactive soils or heighten the extent of moisture-change-induced deformation from active soils. Vegetation has several effects on available soil moisture. In addition to moisture depletion by transpiration, shading of the ground surface, buildup of organic material, retardation of precipitation runoff, and formation of water channels from root disintegration can all influence soil moisture patterns. Large broad-leaf deciduous trees located near structures cause the greatest change in available moisture and the greatest risk of damage to the structure, whether in humid or arid climates. Small trees, bushes, and grasses can affect available moisture at shallow depths, particularly in arid and semi-arid climates. Total vegetation cover, as well as number, size, location, and type of trees affect soil moisture availability. In moisture-accumulation times of the year, the vegetation influence is generally not discernible. However, in moisture-depletion times of the year the influence can be dramatic, as noted in the following case studies.

The contents of this paper are the result of observational studies by the author over the past two years. The case studies evolved from contacts with homeowners or direct observation of local pavement problems. No detailed field investigations were conducted at any of the locations; only general observations or measurements such as surface deformation, crack widths, tree dimensions, etc., were made. Supporting documentation of relevant soil, climate, or vegetation properties were taken from published sources such as USDA Natural Resources Conservation Service (Formerly Soil Conservation Service), Oklahoma Climatological Survey, and Oklahoma Department of Agriculture Forestry Services.

Case 1, 32nd Street, Stillwater, OK

32nd Street establishes a portion of the southern boundary of Stillwater and is located in gently rolling upland terrain. The specific observation site, a section of asphalt pavement, is located in a rural portion of town on the top of a low hill. The site is well drained with drainage to the west and north away from the site. The roadway consists of a gravelly clay layer created by several applications (75 to 100 mm thick) of crusher-run limestone gravel which was "compacted" into the native clay by traffic. After establishing a firm roadbed by traffic, the roadway section was shaped and 75 mm of hot mix asphalt surface course was placed in about 1997. The native subgrade materials are silty clays and clays referred to locally as the Granola clay loam (USDA-NRCS, 1987). Typical properties of the Granola soils are:

Depth (mm)	%-200	% Clay	LL	PI	Moist Density (g/cm^3)	Available Moisture Capacity (mm/mm)
0-150	60-95	27-35	37-50	14-25	1.30-1.60	0.15-0.22
150-475	60-90	35-60	41-70	20-40	1.35-1.65	0.10-0.20
475-875	55-90	35-60	41-70	20-40	1.35-1.65	0.02-0.20

The soils classify as CL or CH in the Unified Soil Classification System (USCS) and A-6 or A-7 in the American Association of State Highway and Transportation Organization (AASHTO) System.

No problems with the roadway pavement were noted until late Summer 1999 when some arc-shaped hairline cracks were noted near the shoulder of the westbound lane near a large Chinese Elm tree growing at the ROW boundary. In late Summer 2000, the cracks had become more obvious. The arc shape of the crack matched the limits of the tree canopy. Figure 3 shows the crack pattern in the roadway pavement, the relative deformation at the cracks, and proximity of the crack to the Chinese Elm tree at the edge of the ROW. The average crack width is approximately 3 mm with no appreciable vertical displacement across the crack. The most likely cause of the crack is shrinkage of the subgrade soil caused by differential drying from the influence of the Chinese Elm tree as it disproportionately depleted the subgrade moisture during the hot, drier climatic conditions.

Case 2, Husband Street, Stillwater, OK

Husband Street, located in the northern part of Stillwater, is a 150-mm portland cement concrete (PCC) pavement resting directly on compacted silty clay and clay subgrade. The subgrade materials were identified locally as the Kirkland silt loam (USDA-NRCS, 1987). Typical properties of Kirkland soils are:

Depth (mm)	%-200	% Clay	LL	PI	Moist Density (g/cm^3)	Available Moisture Capacity (mm/mm)
0-225	80-97	13-26	22-30	2-10	1.30-1.50	0.16-0.24
225-850	88-99	40-60	41-65	18-38	1.35-1.60	0.10-0.14
850-2250	76-99	35-60	37-65	15-38	1.40-1.65	0.10-0.18

The street is approximately 20 years old and had not shown significant cracking until 1994 when some cracking was noted after a small lake approximately 30 m west of the street had been drained for a number of years to reconstruct the dam and spillway. Until draining of the lake (≈1989) and refilling of the lake (≈1994), subgrade moisture conditions were relatively stable because of the shallow ground water table influenced by the lake level. Early in the street's design life, a developer constructed a housing subdivision east of the street and planted numerous Chinese Elm trees along the east side of the street. After refilling of the lake, the city planted numerous Bradford Pear trees along the west side of the street. The cracking problem was exacerbated by the dry seasons in 1999 and 2000. Figure 4 shows the crack pattern along both edges of Husband Street. The cracks are 5 to 10 mm wide with minimal vertical displacement. The cracked portions of the pavement have tilted downward at the edge. The development and perpetuation of the cracks in the pavement are the result of shrinkage of subgrade soils and loss of subgrade support. Both climate and the presence and proximity of trees with higher water demands have obviously made the problem worse.

a. Photograph looking west showing cracks in westbound lane and proximity of Chinese Elm tree

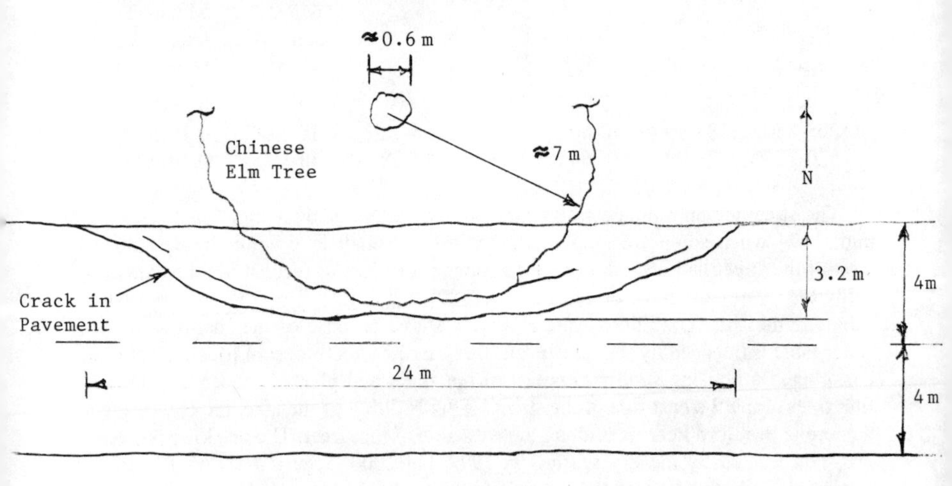

b. Sketch of cracks and proximity of Chinese Elm tree

Figure 3. Arc-Shaped Cracks in HMA Pavement Along 32nd Street, Stillwater, OK

a. Looking north along west curb

b. Looking north along east curb

c. Looking north at both curbs

Figure 4. Edge Cracking of PCC Pavement due to Trees Along Husband Street

Case 3, Residence, Kingfisher, OK

The residence has a perimeter footing with slab-on-grade with minimal steel wire mesh reinforcement. It is one-story with wood frame and brick veneer exterior. The lot slopes generally to the west and north with sufficient gradient to remove surface runoff water. The foundation soils are silty clays and clays identified locally as the Renfrow clay loam (USDA-NRCS, 1984). Typical properties of the Renfrow clay are:

Depth	%-200	% Clay	LL	PI	SL
200-375 mm	99	25	29	9	17
375-750 mm	99	57	58	33	13
750-1000 mm	94	48	47	26	13

The upper soils classify as CL in the USCS and A-4 in the AASHTO system; the lower soils classify as CH in the USCS and A-7-6 in the AASHTO system.

The residence was constructed in 1983 and for nearly its entire performance life showed no signs of distress. In about 1995, the homeowner planted a Bradford Pear tree approximately 3 m from the northwest corner of the garage. By Summer 2000, the homeowner noted a vertical crack in the brick veneer that extended the full height of the wall and through both brick and mortar (see Figure 5). The crack was wider at the top of the wall than at the bottom. Following consultation the homeowner removed the tree and most of the root system. Based on observation records at the time, the crack developed all at once, with the most plausible explanation for the crack being loss of support along the north portion of the west wall due to moisture depletion and shrinkage of the supporting soils. The climatic extremes of Summer 2000 and the proximity of a high water demand tree combined to influence the development of the crack.

Evaluation of Case Studies

In the absence of detailed investigative data, observational and published information was used to assess the conditions and outcomes for the three cases described. These three cases are one-third of the local sites observed in the Central Oklahoma area over the past two years. The cumulative information collected from all the sites influenced the following discussions.

Extremes in climatic conditions, particularly low precipitation combined with above normal temperatures and low relative humidities, significantly contribute to the influence of vegetation on the performance of soils with moderate to high shrink/swell potential. At several of the sites observed, climatic extremes played a major role in causing and exacerbating damage to pavements and lightly-loaded structures. The type and proximity of vegetation interacts with climatic extremes to heighten the problem. In all cases observed, medium to large, broad-leaf, thick canopy, shallow-spreading root trees in close proximity to structures (i.e., Chinese Elm and Bradford Pear) either initiated or worsened the damage caused by shrinking soils. These types of trees cause the greatest influence on the subsurface moisture regime. Experience

a. Close-up view of crack through brick veneer (note: crack highlighted for claritv)

b. Proximity of crack to northwest corner of residence
(Note: crack highlighted for clarity)

Figure 5. Crack in Residence Wall and Proximity of Bradford Pear Tree
(Note: Tree Was Removed by Homeowner After Crack Developed)

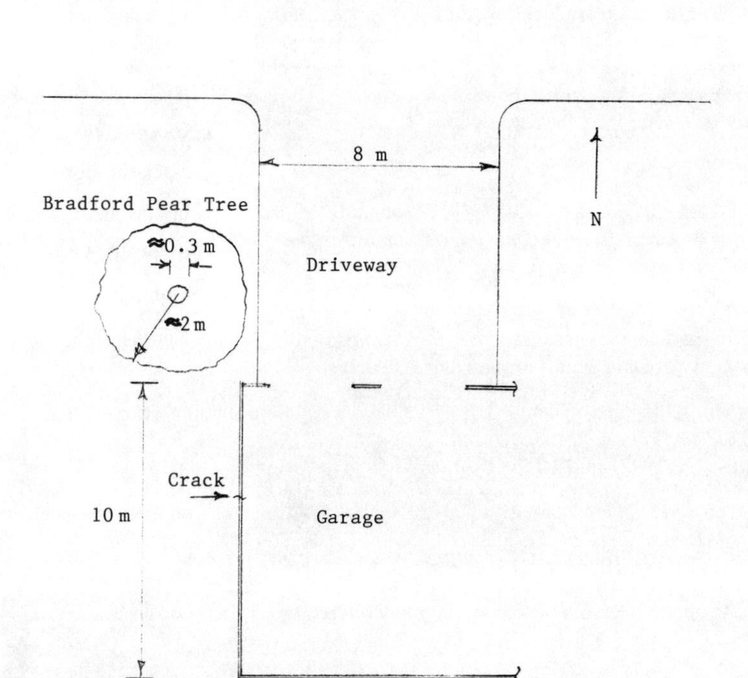

c. Sketch of crack and proximity to Bradford Pear tree

Figure 5. Continued

and observations show that these types of trees should be planted at 0.5 to 1 m beyond the anticipated mature drip line or the anticipated mature height of the tree from pavements or building foundations.

Soil conditions should be considered in landscaping decisions (type and location of trees), particularly with soil having LL > 40 and PI > 25. For landscaping decisions, published information from sources such as USDA Natural Resource Conservation Service county soil surveys and national or local geological surveys are sufficient to address such planting decisions. For local (i.e., city or county) pavement subgrades, shrink/swell potential can be effectively and economically addressed by chemical modification (lime, fly ash, cement kiln dust). In addition to reducing the shrink/swell potential of the treated layer, it places the pavement structure farther from the active soil subgrade.

References

Haliburton, T.A. and Marks, B.D., "Subgrade Moisture Variations in Expansive Soils," *Proceedings of Second International Research and Engineering Conference on Expansive Clay Soils*, Texas A&M University, College Station, TX, 1969, pp. 291-307.

Oklahoma Climatological Survey, *Oklahoma Monthly Summary*, December 2000, University of Oklahoma, Norman, OK, 2000.

Snethen, D.R. and Huang, G., "Evaluation of Soil-Suction Heave Prediction Methods," *Proceedings of Seventh International Conference on Expansive Soils*, Dallas, TX, 1992, pp. 12-17.

Snethen, D.R., "Three Case Studies of Damage to Structures Founded on Expansive Soils," *Proceedings of Fifth International Expansive Soil Conference*, Adelaide, South Australia, 1984, pp. 218-221.

USDA Natural Resource Conservation Service, *Payne County Soil Survey*, April 1987.

USDA Natural Resource Conservation Service, *Kingfisher County Soil Survey*, May 1984.

Expansive clay problems - How are they dealt with outside the US?

Richard Radevsky[1]

Abstract

People in countries all over the world have to deal with houses built on expansive clays (clay soils that shrink and swell). The ways in which they tackle the problem vary considerably. Not only do solutions differ technically but they also depend on the property market, the legal framework of the country, insurance policies, the attitude of insurers, the history of the region, the degree of building regulation, the experience and training of construction professionals and perhaps above all the sensitivity of the house owner to the cracks in his house. Relatively minor cracking is seen in some places as an early sign of impending disaster. Rarely is the house owner's attitude in accordance with what technical research tells us is the structural significance of cracks.

In the UK with the involvement of insurers for nearly three decades, there is high sensitivity to extremely small cracks. Costly procedures are used to deal with them. In Australia, South Africa and USA foundations consist mostly of reinforced concrete rafts which are cheap and quick to underpin. Where insurers are not involved house owners pay for their own remedial works, sometimes using additional mortgages. In general it appears that private financing and little local authority regulation of underpinning produces low cost and variable quality solutions.

Introduction

Expansive clay problems are experienced in many countries throughout the world. These include U.S.A., Canada, France, Spain, Denmark, South Africa, Australia, Israel, Romania, China, Saudi Arabia, Zimbabwe and the U.K. Approximately

---

[1] EUR ING RICHARD RADEVSKY BSc CEng PEng FICE, MCIWEM MinstPet, MIFireE FCIArb, Principal, Charles Taylor Consulting plc, London, UK; Tel +44 20 7488 3494; Fax +44 20 7488 7544; RARadevsky@cs.com; www.mutuality.com

one quarter of the surface area of the United States is particularly affected, and this includes every state (Simons 1991). A publication by American Society of Civil Engineers (American Society of Civil Engineers 1995) states that the effect of expansive soil damage on a local, regional and national scale was considerable. As early as 1973 (Jones and Holtz 1973) the annual cost of expansive soil damage in the U.S.A. was estimated at $2.2 billion, excluding that caused by earthquakes, hurricanes and floods combined in an average year. In 1980 (Krohn and Slosson 1980) , the annual cost of expansive soil damage in the U.S.A. was estimated to be $7 billion. The damage, however, is not limited to buildings with shallow foundations. It includes roads, services and other structures, including commercial buildings.

This papers concentrates of the effects of expansive soil damage on buildings with shallow foundations, chiefly private houses.

Recently a study comparing five countries, which have similar soil conditions, buildings, legal system, culture and level of wealth, use of mortgages and use of insurance was carried out for insurers in UK (Radevsky 2000). The countries studied were the U.K., South Africa, Australia, France and the U.S.A. (particularly Texas).

House Construction

U.S.A.

Houses in Texas tend to be similar to those in Australia, in that a large proportion is single storey detached houses, built with brick veneer construction. Internal wall surfaces tend to be plasterboard with a skim coating. The design of houses in some areas is more elaborate than would typically be seen in Australia with more bedrooms and other features. There are some semi-detached houses and some maisonettes and few areas of older masonry houses.

U.K.

Residential buildings in the U.K. are mostly of two or three storeys and are constructed of brick/blocks with pitched tile/slated roofs. High housing density in main cities is common with many semi-detached and terraced houses. The housing stock is older than in many countries with a large number of houses more than 50 years old. A particular feature of U.K. houses is the use of wet plaster finishes applied directly to masonry walls. This coats the walls with a brittle surface, which is particularly susceptible to formation of cracks as a result of small movements.

South Africa

In South Africa, single storey houses account for perhaps 80% of all houses. The amount of land available for housing means that most houses are detached. Roofs are often of lightweight metal sheeting, although tiled roofs are also common. Walls tend to be of brick and wet plastering is the most common finish. External render is also widely used. A typical South African house would be a large detached bungalow.

Australia

80% of all houses in Australia are single storey. The land available for housing means that most houses are detached. Semi-detached houses and two storey houses do occur but in much smaller numbers than in the UK. Older houses in one of the cities studied, Adelaide, are commonly constructed with stone external walls and wet plastered internal finishes. Roofs are of corrugated metal sheeting or tiles. Internal floors are generally of suspended timber construction. Some houses have solid concrete floors, particularly in the "wet" areas of the house, i.e. the bathroom, toilet and laundry rooms. Brick veneer construction is common where houses have an outer skin of brickwork and a timber internal framework covered with plasterboard. Wet plastered internal finishes are very uncommon. The structure of the building is held together by a timber framework and a brick veneer is for waterproofing and decorative purposes. Brick veneer construction was developed in Australia. Some lightweight structural frames are beginning to be used instead of timber.

France

Buildings in France tend to be of a very mixed nature. There are single storey, two, three and four storey buildings, which are detached, semi-detached and terraced. There are a large number of old properties, including stone houses.

Foundation Design

U.S.A.

Foundations particularly in Texas, are generally in accordance with the Uniform Building Code, which requires an engineer or architect to design them. Slab-on-grade and raft foundations typically have an edge beam extending to a depth between 450 (1'6") and 750 mm (2'6"). Reinforcement consists of post-tensioned cables or reinforcing steel bars. Slab on grade, i.e. not significantly penetrating into the ground, is used even for some commercial buildings.

Many house foundations are constructed on soils without any form of pre-treatment. Pre-treatment prior to construction is carried out often on commercial buildings and a variety of techniques are used.

U.K.

The majority of houses in Britain are built with deep (1m to 3m – approximately 3' to 10') unreinforced concrete strip footings. Very old houses may have corbelled brickwork footings. The use of reinforced concrete rafts is very unusual. Most new houses comply with standards set down by the National House Building Council, which provides 10 year defects guarantees. The standard set down by the NHBC is generally more stringent than is required under the statutory Building Regulations.

South Africa

South Africa is generally not exposed to significant frost and foundation depth normally extends to only 500 mm (approximately 1'8") below ground level without reinforcement. Swimming pools are common in South Africa and are often built close to houses. South Africa's National House Building Registration Council provides guarantees against defects found in new houses in the first 5 years. The NHRC classifies sites according to the type of foundation material specified and expected ranges of total soil movement. Generally, where movements of less than 7.5 mm (approximately 0.3") are anticipated, a competent person is not expected to carry out foundation designs. New building works are subject to some controls by local authorities. Reinforced concrete raft foundations are widely used for new houses, including waffle slabs.

Australia

Most houses have very shallow foundations, often less than 500 mm (1'8") below ground level. A typical concrete strip footing used, was traditionally 450mm wide and 300 mm (1') thick, founded at a depth of 450 mm (1'6"). These used to be hand dug and, as a result, were often wider than the minimum requirement. With the introduction of backhoes in the 1960's, footings were excavated to precisely the minimum width and have been found to be more susceptible to foundation movements than those constructed earlier. They remained unreinforced. In 1981/83, droughts caused widespread cracking in houses and the design of footings was re-evaluated. As a result, slabs with a 450 mm (1'6") deep edge beam became standard and soil testing became compulsory.

A raft is now the most common form of foundation for new houses. Some slab foundations are founded at ground level. A contractor can only build new houses after a soil investigation and design by an engineer.

France

Foundations for houses tend to be un-reinforced strip foundations of the same type that are used in the U.K. Foundations commonly only extend to a depth of 0.8 m (2'8"), the depth having being selected to avoid exposure of the foundation sub-soil to frost.

Soils

U.S.A.

The soil in large areas of the U.S.A. is expansive. Texas is often regarded as the area of the U.S.A. with the most extensive and severe expansive soil problems. The problems emanate from the considerable difference in climate than occurs seasonally. Winters tend to be wet and cool and summers extremely hot and dry. Severe droughts have occurred in Texas in the recent past. Clay soils in the area swell and shrink considerably owing to changes in moisture caused by a combination in the weather, the effect of trees close to properties and the release of water into the soil, either by watering or by leakage from pipes and sewers. Different strata of clay shrink and swell at different rates. In addition, the moisture of the soil at the time of construction can make a considerable difference in the way that the foundation will behave thereafter. A house built during a drought, when the foundation soil is dry will behave differently from a house built during a season when the soil is moist.

Distortion of the reinforced concrete raft is the most common form of damage, which can cause changes in levels of 10's or even 100's of millimetres. Such movement can cause slabs to tip, crack or distort.

U.K.

The majority of U.K. expansive soil problems, (known in the U.K. as Subsidence), result from the shrinkage of saturated clay soils caused by drying. The drying is most often caused by a combination of dry weather and the extraction of water from the soil by the roots of trees and shrubs close to the buildings. Less common are problems of softening soils by drain or water main leakage, heave and mining. Expansive soils are found in a wide area across the U.K., with the highest concentrations being in the south-east portion of the country.

South Africa

Soil problems in South Africa can generally be divided into three categories. Expansive clay, collapsing sand and dolomite sink holes.

Expansive clay is similar in nature to the shrinkable clays widely present in the U.K., but for the most part they tend to be in a desiccated, (shrunk) condition and cause damage when they absorb water and swell. In some areas, properties also suffer from drought/tree related clay shrinkage where the moisture content of the clay soil is reduced. Expansive soils can be found in strata up to several tens of metres thick. Heave up to 200 mm (8") can occur naturally. Heave over 750 mm (2'6") has been seen under experimental conditions.

Collapsing sand is loosely cemented sand of weathered granite, which tends to become unstable when exposed to excessive water. Such sands are often interspersed with hard rock formations, making properties highly susceptible to differential movements.

Dolomite is a soil with a structure that is susceptible to the development of sinkholes. Sink holes usually form when water runs through the soil to a low-level void. After some time large holes can suddenly appear at the surface.

Australia

Certain parts of Australia, particularly Adelaide, Melbourne and Brisbane, have expansive soils. These soils produce both heave, when subjected to increased moisture content, and shrinkage induced subsidence, when their moisture contents are reduced. There are also collapsing sands in southern South Australia, caused by changes in the chemical composition of the soil, but this problem is far less severe than the expansive soil problem. It is common for houses to be constructed during the dry months and for the soil to expand later as it becomes wetter. In Adelaide, it has been estimated that 80% of damaged houses were damaged as a result of the action of trees. In Melbourne, many streets are lined with well-developed Plane trees, which have caused substantial damage not only to streets, but to nearby houses as well.

France

Soil problems found in France are very similar to those found in the U.K. There are large areas where properties are built on shallow un-reinforced foundations on shrinkable clay. There are four main areas of France that produce the most losses as follows:-

- The north of France where peaty clays cause problems with pronounced shrinkage movement and significant cracking;
- The Loire area near Tours and Orleans;
- The suburbs of Paris
- The south around Toulouse.

Problems in the north of France are such that owners sometimes face difficulty selling their property, because of the frequency of damage even with subsidence insurance.

Insurance Policies

U.S.A.

Insurance in Texas is heavily regulated. Cover is generally not available for damage associated with foundation movement unless it can be shown to be the result of water escaping from underground drains. Currently insurers face a large number of disputes over whether movement to houses was or was not associated with a leak. Older houses tend to have cast iron drain pipes, (rather than p.v.c.), and tend to have heave problems more often than others.

U.K.

Since 1971 most U.K. household insurance policies have included cover for subsidence, landslip and heave. Most policies are subject to £1,000 excess, raised from £500 in the relatively recent past. Policies generally exclude coastal erosion, design defects, damage while the building is undergoing structural alteration, damage to solid floors unless the walls are damaged at the same time, and damage which is covered under guarantees.

House owners are generally asked to declare at the time that they take out their policy whether their house is built in an area where there has been evidence of damage by subsidence.

Insurers have relatively sophisticated computer-based subsidence underwriting tools available to them, these include data on soil and claims history tied to postcodes. They are able to identify the relative subsidence risk applying to relatively small groups of properties.

South Africa

Subsidence losses are generally not a major issue for insurers in South Africa. This is largely because insurers do not pay large numbers of claims for structural damage to houses caused by foundation movements. Proposal forms for insurance do not include questions related to soil conditions or foundation movement and not all insurance policies provide cover for subsidence or even for the foundations of the houses. Most claims for foundation related damage are rejected by most insurers. A substantial proportion are rejected on the basis that the damage has resulted from defective construction of the property insured, (i.e. that the foundations were inappropriate for the soil conditions and in an area of the property). Of those claims that are accepted, insurers normally only pay for the repairs to the superstructure, not for any underpinning work that is required. This is on the basis that underpinning is a preventative measure.

Australia

Household insurance policies generally do not include subsidence or heave as an insured peril, except with very restricted cover. The cover for subsidence that is provided is only where the cause of the subsidence is tied to another peril, such as storm, rainwater, wind or earthquake. There tend to be a large number of claims for damage to houses resulting from breakage in underground water pipes. In general, foundation movement is not considered to be a major problem for Australian insurers.

France

95% of the French population has insurance. This includes natural perils insurance. Such policies have traditionally dealt with events such as floods and earthquakes, but in 1989 the French government declared a natural catastrophe event to include a drought which caused subsidence. This produced huge numbers of claims.

The cost of natural catastrophe claims is financed by tax until recently at the rate of 9% on all insurance premiums paid. Insurers collect the tax and are then able to either retain the tax, in which case they must finance all natural perils losses or they can pass up to 90% of the tax to the state reinsurer in return for corresponding reinsurance coverage and stop loss cover. The state reinsurer suffered such large losses as a result of the 1989 and 1996 events that it had to apply to the French government for an increase in the tax rate from 9% to 12%.

Differential premium rating of properties in areas known to be prone to foundation movement is not permitted. Insurers generally do not collect information at proposal stage concerning the construction of the houses, their foundations or the proximity of trees.

Mortgage Lenders

U.S.A.

Mortgage companies in Texas are generally not concerned about the susceptibility of properties to foundation movements, partly because new houses are generally subject to 10 year warranties and partly because the cost of remedial works is not particularly high in relation to the relatively high value of the house.

U.K.

Most houses in Britain are purchased with the benefit of a mortgage given by large financial institutions including banks and building societies. Mortgage lenders generally rely on insurers to provide insurance coverage for foundation problems. If insurance coverage is not provided then mortgages are generally not available. The mortgage companies therefore reply on insurance companies to determine which properties are suitable for insurance and which are not. Currently, mortgage lenders do not generally become involved in the susceptibility of houses to foundation movements, assuming that the insurance company will deal with any remedial works that are required.

South Africa

Mortgage lenders in South Africa generally do not concern themselves with foundation problems. The majority of the cost of foundation remedial works is privately financed and homeowners will frequently have to take out a second mortgage on their house to pay for the works. With a typical interest rate of 19%, a second mortgage can impose a significant financial burden on a homeowner. Mortgage companies generally have houses valued before they provide mortgages, but such valuations are more focussed on the resale value of the property rather than any technical defects.

Australia

Banks in Australia provide mortgages, frequently up to a very high proportion of the value of the property. The subject of foundation movement is generally not something which Australian mortgage lenders are concerned about, including the general ability for houses to be sold even where they are subject to cracking.

France

Mortgage companies are generally not concerned about the susceptibility of houses to foundation movement owing to the fact that universal insurance for such problems is available to deal with foundation remedial costs.

Remediation Techniques

U.S.A.

Houses in Texas are generally founded from reinforced concrete rafts. These rafts are used as pile caps and a large number of remedial techniques have been developed (Foundation Performance Committee 1998) including: Root Shields, Perimeter Watering, Stabilisers, Foam Injections, Mud Jacking, Spread Footings, Block and Base, Bellbottom Piers, Straight Shaft Piers, Builders Piers, Tunnelling,

Solid Pre Cast Piles, Pre Cast Piles, Reinforced Steel Pipe Piles and Helical Piers. The cost of remedial piling is substantially lower than in the U.K. and the time required to complete the operation is very short indeed, running only to a few days. Some irrigation systems have been used, however these can be subject to water restrictions during a drought which is often just at the period when the irrigation systems are most needed.

U.K.

Underpinning is the most well known method in the U.K. for the treatment of subsidence damage. This involves installing additional deep unreinforced strip footings beneath the original foundations of the house that is damaged. In addition to unreinforced strip footings, other more sophisticated techniques involving the installation of piles and micro-piles are used. There is little use of soil grouting because of the very low permeability of the clay soil, which is involved in most cases. Rehydration techniques are not widely used. There is little use of jacking to return properties to their original elevation, mostly because the extent of vertical movement is relatively small. A considerable amount of time, effort and money has been invested in the development of a small piling rig capable of passing through a single doorway so that they can carry out piling work from inside a house without extensive demolition. Many remedial techniques developed in USA and Australia are not suitable for use on UK house foundations since they rely on the existing house foundation to act as a pile cap. As UK house foundations are generally unreinforced they are incapable of withstanding point loads without a pile cap being installed beneath them first.

Between 1971 and the early 1990's a large amount of money was spent by insurers on remedial techniques, particularly underpinning. From the early 1990's onwards there has been a general move towards treating underpinning as a solution of last resort and there has been greater use of monitoring of movement once the cause of the subsidence, (often trees or shrubs), has been neutralised.

South Africa

In most cases where subsidence is experienced by a house, the cost is met by the houseowner. Underpinning does not require local authority approval. Where reinforced concrete raft foundations are present, piling using the raft as a pile cap has been used.

One of the more common techniques is to install articulation joints in buildings to allow them to flex rather than attempting to prevent the house from moving further. Post construction articulation requires large cutting equipment to install joints in houses, the joints being covered with flexible strips. In some cases, joints are found to open during construction as stresses are relieved. The presence of joints in a house does not appear to deter sales. Sometimes recommendations for watering are given at the time that the joints are installed. Houseowners are recommended to watch for the joints in the house opening or to watch drought sensitive plants to indicate when additional watering is needed. Problems have occasionally occurred where the houses have been rented and the tenants have neglected to water the plants, or else the owners are nervous of the responsibility to water appropriately.

Australia

The Australian Building Code No: 2870 states that

> *"Underpinning should generally be avoided where the problem is related to reactive clays".*

This is because the depth of the expansive soil is mostly such that underpinning is not considered a technically appropriate solution. Irrigation systems are relatively frequently used. Owners however do not often wish to wait 12-18 months for irrigation systems to rectify the movement of their house. They often ask for underpinning which is privately financed in order to produce a quick fix. They will then repair cracks themselves at the property. Reports of dramatic crack closing resulting from watering of houses have been made, even where cracks have closed from 22 mm (0.9") wide to 7 mm (0.28") wide in the space of a week. Consulting engineers and architects working in major expansive clay areas of Australia report that Australians are prepared to live with cracks in old houses that, from the author's experience, would not be tolerated in UK (Radevsky 2000). Cracks up to 10mm wide are often regarded as normal and regular repair is a maintenance matter. Solutions such as the irrigation of soils to promote recovery generally only eliminate the most severe damage. Commonly this leads the building to be repaired with new cracks up to 5 mm (0.2") occurring regularly each seasonal cycle.

Irrigation systems commonly consist of rock filled bore holes or slotted pipes packed with Geotextile fabric with internal automated drip water feed. In some cases, tree root repellent chemicals have been added around the irrigation system to deter root growth.

France

A variety of techniques for remedial works are used in France, including conventional underpinning, micro piling and the use of irrigation systems. Trials of new methods of remedial works are currently underway, some involving the minimisation of evaporation to retain water within the ground. Tree control is not as frequently used as in the U.K. for a variety of legal and cultural reasons.

Pre-Purchase Condition Surveys

U.S.A.

In general, people do not have pre-purchase surveys, (of the same type as are carried out in the U.K.), they have inspections by warranty company inspectors. The warranty companies provide cover for appliances. Inspections by these companies may identify cracks in the building and may suggest that an engineer inspect the damage to the house. There are companies specialising in property inspections. Professional engineers occasionally carry out inspections. Mortgage companies do not generally worry about soil conditions or the susceptibility of houses to foundation movement.

U.K.

Chartered Surveyors, trained to identify signs of previous subsidence, or evidence of current foundation movement survey most houses at the time of re-mortgage. This does not prevent occasional mistakes or mis-diagnosis of problems (Radevsky 1993), but it does result in the identification of large numbers of subsidence damaged houses which might otherwise go undetected for a considerable period. Most mortgage lenders require pre-purchase surveys to be carried out. Signs of subsidence identified by a pre-purchase surveyor can kill a house sale and adversely affect the value of the house until the problem has been rectified.

South Africa

Most houses in South Africa are the subject to bonds, (mortgages), with banks. Before banks provide the bond, the property concerned has to be valued by a valuer usually from the staff at the bank. Such valuers generally restrict their work to the market value of the property in relation to the security the property has to provide against the bond. Bank valuers are generally not qualified surveyors. South Africa does have a separate professional body of Chartered Building Surveyors, as in the case in the U.K. These are registered professionals but they are not normally involved in house surveys.

Australia

Banks providing mortgages carry out pre-purchase valuations often using their own staff who are trained to recognise the presence of building defects, but not necessarily to evaluate their significance. If a defect is identified by a valuer's inspection, either the mortgage will be refused or the buyer of the property will be advised that they must obtain a report from someone licensed to carry out building inspections. If the inspection report identifies a significant defect, either the mortgage will be refused or a condition of the mortgage will be the rectification of the defect. The bank may withhold part of the mortgage money until the defects are rectified and may inspect the building again during the works. The bank's valuation report is not released to the potential homeowner and they are advised to obtain their own valuation from an independent valuer. Owing to the cost of such independent valuations, many potential homeowners do not obtain their own reports. Many homes are sold by auction without surveys. A division of the Royal Australian Institute of Architects has developed a pre-purchase survey system and this scheme has been growing in significance. The survey service is relatively cheap and is used by many architects as a method of attracting new clients. Whilst identifying defects, the report also covers the renovation and development potential of properties and architects usually try to meet their client for whom they are carrying out the survey. Unlike in the U.K. where pre-purchase surveys tend to emphasise the negatives, the Australian survey reports balance defects with ideas for improving properties.

France

Most people in France do not bother with pre-purchase survey reports. Occasionally, they may consult an architect who is a friend to look at a house they are thinking of buying. His views may be used as a means of negotiating a reduction in price. Written reports are very rare. The current system means that all houses in France are insurable for natural catastrophe insurance coverage, (including subsidence cover). Lending institutions are therefore not concerned about the possibility that the house would be uninsurable.

New House Guarantees

U.S.A.

Many mortgages in Texas are arranged for first time buyers through the Federal Housing Administration, (FHA), at advantageous interest rates. Houses built for FHA financed buyers and some financed by certain mortgage companies have to be provided with 10 year warranties and the houses have to be designed and constructed to more rigorous standards than other houses. Soil investigations and engineering involvement in design and construction supervision are required in areas with known soil problems. The warranty covers the house buyer. The largest warranty company in Texas, Homebuyers Warranty, views construction on fill as a more serious problem than expansive clay; Texas has more claims than any other state, the claims loss ratio has dropped recently due to tightened control on building methods. The company also has approved engineers for design work and is reinsured internationally.

New houses are normally provided with a 1-year builder's warranty. Under Texan law the builder is also liable under an implied warranty for 10 years for defects but many Houseowners are not aware of this. For a warranted claim to succeed, the level of damage to the house has to be relatively high.

U.K.

Although buildings in the U.K. have to be subject to planning and building control approval, most new houses are also built with a 10 year defects guarantees provided by the National House Building Council, (NHBC). The NHBC sets out standards for the nature and extent of the foundations required to new houses, which are far more stringent than building regulations with regard to the depth of foundations in relation to trees and tree roots. It is therefore often the NHBC regulations, which tend to govern the design of foundations in areas where there is shrinkable clay sub-soil.

South Africa

South Africa's National House Building Regulation Council, (NHBRC), provides guarantees against defects found in new houses in the first 5 years. The Council classifies sites according to type of foundation material and specifies an expected range of total soil movement. Generally, where movements of less than 7.5 mm (0.3") are anticipated a competent person is not required to carry out foundation designs. The definition of a competent person is someone who is professionally qualified and able to obtain professional indemnity insurance to the value of at least approximately US$70,000. Currently, lending institutions will mostly refuse to provide loans on a new house without a guarantee being in place. Legislation is shortly to be enacted making it compulsory for all new houses to be the subject of NHBRC guarantees. If defects do develop, the cost of rectifying them will be paid for by the NHBRC if the builders do not rectify the defects. NHBRC are able to recover their losses from the builder who constructed the house, or the structural engineer who designed it if negligence is proven. Currently there are around 80,000 policies in force in South Africa and only a handful of claims relating to foundation movements. NHBRC has its own home builders manual for house constructions, including foundation design codes produced by the professional engineering institutions in South Africa.

An alternative to the NHBRC scheme is operated by Homesure to provide a similar new house warranty but is prepared to continue cover almost indefinitely provided the houseowner continues to pay premiums. The policy effectively changes from the new house warranty into a house buildings policy as time progresses. Such policies do provide cover for ground movement, including subsidence and heave.

Australia

By law all new buildings costing more than US$3,500 have to have homeowners warranty indemnity insurance policy. A typical policy will cost a builder US$400-US$450 per house and covers up to US$55,000. In some cases, the limit is soon to be raised to $100,000. To make a claim costs the houseowner approximately US$300, which is refunded if the claim is valid. In Adelaide, the policy includes a clause excluding the cost of rectifying damage related to soil movements. New house guarantee schemes vary from state to state. Many states used to have state administered schemes although most have now been privatised, for example, Victoria's Housing Guarantee Fund is in run off. In Queensland the Building Services Authority maintains a state wide guarantee monopoly. It covers all new houses built by a licensed builder for 6½ years. The scheme covers 45,000-50,000 houses per year and pays out around US$2 million on foundation problems on new houses. The scheme is internationally reinsured and in other states new house guarantee insurance policy conditions vary considerably. The Building Services Authority have sought through research to highlight the most common factors in foundation problems of houses under 6½ years old (Building Services Authority 1998). This highlighted higher than normal rates of failure in houses built by certain builders and under the supervision of certain engineers.

New houses are frequently supplied with a pamphlet produced by CSIRO, the Australian building research organisation, explaining how the foundations of houses are expected to perform and what measures are needed to care for the foundations (CSIRO Australia 1996).

Tree Root Liability

U.S.A.

In Texas, most street trees are planted by the city in the central reservation of roads rather than in verges. Houseowners are generally responsible for maintaining verges down to the kerb including trees within the verge. Investigations did not uncover any case where an individual has been sued or a city has been sued for tree root damage. Action against neighbours for tree root trespass appears generally not to be pursued.

U.K.

The U.K. courts have been developing case law related to tree root trespass related to both private individuals and local authorities who are responsible for the large number of street trees found in many subsidence prone areas. The proliferation of tree root trespass claims caused the Association of British Insurers to formulate an agreement for its members not to pursue most claims against private individuals. The tree root trespass position however results in many trees being reduced and allows insurers to recover the cost of damage to houses where trees are the responsibility of the local authority.

South Africa

The subject of tree root liability is not one, which appears to be well developed in the South African legal system.

Australia

Local authorities in Australia are conscious of liabilities associated with street and public trees. One authority for example received 49 claims during one year, June 1997 to June 1998. They generally take steps to control risk. Liability can arise either as a result of damage or on the grounds that there has been a negligent selection of the wrong variety tree in a particular location, (i.e. of a variety with a high water demand and aggressive water searching characteristics). Some local authorities believe in root barriers to prevent losses and have been researching into appropriate specifications for the size, composition and optimum location of such barriers. Local authorities occasionally pay for irrigation systems for street trees. In the longer term, very high water demand trees are being eliminated close to houses. Typically, a Eucalyptus tree may consume 200 to 250 litres of water per day. Some Australian trees are known to affect properties three times their height away (Sheard and Bowman 1996).

France

The French legal system does not provide legal grounds for a houseowner to seek damages from a neighbour where a tree's roots are causing subsidence. Under the French Civil Code, a person can only complain about a neighbour's tree if it is 2 metres or less from the boundary. Local authorities often do not respond positively to requests for tree reduction or removal. Many French properties are part of co-operative estates and have requirements for each lot to contain a minimum number of trees. Requests by insurers for tree removal are sometimes treated with a refusal on the basis that the co-operative estate rules require the trees to be there.

In practice, most people who are shown that their tree is causing damage to their house will reduce the tree. If a neighbour refuses to reduce a tree, there is effectively nothing that can be done.

Expectation of House Owners

U.S.A.

The view of many people in Texas is that they live for today and are constantly looking to move onwards and upwards. They move house every few years and do not view a house as something, which is expected to last forever. There are a few historic buildings, which are preserved, but most buildings are very recent by European standards. Older houses built 40 or 50 years ago would often be viewed as re-development opportunities bought for their plot value. The house would be demolished and replaced with something larger. The view of a house as something, which is to be lived in for only a few years means that there is less concern over the durability of the repair techniques than is the case in the U.K. Provided a repair will outlast the current owners, and allow them to inform a purchaser that the problem was fixed the owners will be satisfied. Analysis has been carried out to examine how much damage homeowners can tolerate before they felt that action was required (Koenig 1992). This suggested that there was at approximately 1¼" (31 mm) of slab deflection, 10% of foundations would be repaired, at 2¾" (68 mm) of slab deflection 50% of foundations would be repaired and it was not until 4¾" (108 mm) was reached that 90% or more foundations would be repaired. This shows a far greater level of tolerance of movement than in the U.K., however foundations in the U.S.A. tend to be reinforced concrete rafts rather than unreinforced strip footings. With regard to newer houses, there is pressure on builders to keep prices low, particularly at the lower end of the market. It is difficult therefore for house builders to introduce improvements in the quality of their foundations and remain competitive.

U.K.

Owing to the scale of past losses, the U.K. household insurers are very aware of the subsidence risk. Many house owners know about the problem and are sensitive to the appearance of even relatively fine cracks in their house (Biddle 1998). This situation appears to have arisen since subsidence insurance was introduced in 1971. Before that time, cracking in houses was generally tolerated and was something, which a decorator would deal with. Today, cracks in houses can have an adverse effect on the house's saleability so crack repairs and remedial foundation work are widely used at a considerable cost.

South Africa

The level of awareness of foundation problems amongst the general public and even much of the construction industry is low. The problem does not appear to impact significantly on the property market, or be of great concern to insurance companies or banks providing mortgages. The result is that expenditure on repairs on the foundations of damaged houses is low. People generally accept the presence of cracks in their houses. Minor cracks or recently significantly repaired cracks do not generally prevent house sales.

Australia

Australia has never had insurance coverage for subsidence or heave. The absence of the insurance does not generally worry people. Many houses, perhaps most frequently in Adelaide, have cracks and these are accepted, particularly in older stone houses. Older stone houses are generally sought after for their central location and historical connections. Some are heritage listed. Owners accept the presence of large cracks, which in the U.K. would cause considerable disquiet and render a house uninsurable, unmortgagable and unsaleable.

The regular occurrence of new cracks is also widely accepted. Regular crack repairs are considered a normal part of house maintenance. This does not apply in all areas of Australia. In Melbourne for example, cracking does cause concern. People here are considered more legally aware than many in other areas of Australia and they are 8 times more likely to have a pre-purchase survey carried out than those in Sydney, for example. Where a severe crack develops, consulting engineers normally recommend adjustment to the environment around the house. This often involves the removal or reduction of trees, the installation of irrigation and repairs and redirection of the drainage system. Charging methods for water useage are changing from a fixed charge per house to a metered basis. It is feared that those transferring to a metered supply may not run their irrigation systems and subsidence damage may result. A few instances of this have already been seen and there are concerns that such cases may become more common. Significant external cracks are often left untended although it should be remembered that there is no significant frost problem so erosion of the masonry either side of the crack does not normally occur with time.

Cracks are often noted in pre-purchase survey reports and sometimes engineers will be called in to evaluate the significance of larger ones. More often than not, this leads to advice on managing the cracks rather then the recommendations for the house not to be purchased.

France

Owing to the insurance situation, house owners and lending institutions in France do not have to worry that properties will become uninsurable. There appears to be relatively little pressure to improve the resistance of new buildings to subsidence by tightening control on building regulation compliance. The increasing level of losses is stimulating debate about the problem. However, for political reasons charging everyone a uniformly higher premium may be used as a short-term solution. Many French insurers believe that the advice received by house owners from contractors and consultants is over conservative and this is provided in order to minimise the possibility of the contractor or consultant being sued in the future. The consequence of this is that remedial works costs have risen considerably. The cost of remedial work is significantly higher in the suburbs of Paris then in other areas. The increased cost in Paris is accounted for by the severe damage in the Paris area, combined with pressure from people for substantial remedial works, which can add value to a person's house.

Overall Conclusions

Before any meaningful comparisons can be made between different countries, it is necessary to understand certain significant differences. Most houses in South Africa, Australia and Texas are detached and single storey, as compared to the U.K. and France where there are many houses of two and three storeys high and frequently semi-detached or terraced. In the U.K. and France there is an exposure both to clay movement and frost, as a consequence houses here built from the 1960's onwards generally have foundations approaching 1 metre deep. In the U.K. and France there are large numbers of insurance claims accepted by Insurers. In the U.S.A. large numbers of claims are met for foundation movement but under the peril of escape of water since subsidence and heave are not insured perils. The most common form of foundation in Australia and Texas is a reinforced concrete raft, which is inexpensive and quick to underpin.

Where house owners, mortgage lenders and the professionals advising them are not anxious about cracks in houses, the presence of cracks does not affect the property market. In the U.K, apparently unintentionally, a system has developed which has resulted in large expenditure on subsidence remedial works. Where insurers provide cover for subsidence damage, more extensive remedial works tend to take place than where private individuals use their own money. Where an individual pays, there is a greater tolerance towards cracks in house walls. Experience in some countries shows that in the absence of concerns over the saleability of their houses, most house owners appear prepared to undertake occasional crack repairs as part of the normal maintenance. House owners want reassurance that in the event of a serious problem of structural significance requiring high cost foundation works, insurers in countries where insurance coverage is provide will meet the cost.

References

American Society of Civil Engineers (1995) – "So your house in built on Expansive Soils........" Report of Shallow Foundation Committee of the Geotechnical Division, New York N.Y.

Biddle P.G. (1998) - "Tree Root Damage to Buildings" - Willowmead Publishing Ltd., Wantage, UK.

Building Services Authority, Australia (1998) – "Subsidence of residential buildings" – A Case Study

CSIRO Australia 1996 "Guide to home owners on foundation maintenance and footing performance (updated for AS 2870-1996) CSIRO Publishing, PO Box 1139, Collingwood, Vic 3066, Australia.

Foundation Performance Committee (1998) - "Foundation Repair Methods" - 1998 Symposium Building Better Foundations for the Residential Home Industry - Houston Texas - 5th November 1998.

Jones D.E. and Holtz W.G. (1973) "Expansive soils – The Hidden Disaster" Civil Engineering – ASCE Vol. 43 No.8, New York, N.Y. August, pp. 49-51

Koenig A.S. (1992) "Slab-on-Grade Deflection: How much is too much?" - 7th International Conference on Expansive Soils Dallas, Texas, August 3 – 5, 1992, pp. 89-92

Krohn, J.P. and Slosson, J.E. (1980). "Assessment of Expansive Soils in The United States" Proceedings, 4th International Conference on Expansive Soils Denver, CO, Vol. ,1 June, pp. 596-608

Radevsky R. (2000) Subsidence - A Global Perspective - Association of British Insurers - Research Report No. 1, London, England.

Radevsky R. (1993) "Why do Structural Surveyors and Valuers Go Wrong with Subsidence?" - Structural Survey, Volume 11, No.4 (Spring 1993) pp. 354-356

Sheard M.J and Bowman G.M (1996) – "Soils, stratigraphy and engineering geology of near surface materials of the Adelaide Plains" – joint publication of Mines and Energy of South Australia and CSIRO Division of Soils Adelaide South Australia Chapter 7 Environment, Vegetation and Site Management pp. 73-76

Simons, Kenneth B. (1991) American Society of Civil Engineers – "Limitation of Residential Structures on Expansive Soils" Journal of Performance of Constructional Facilities, Vol 5, No. 4 , pp. 258-269

In-Situ Modification of Active Clays for Shallow Foundation Remediation

Arthur Pengelly[1] and Marshall Addison[2], M.ASCE

Abstract

Damage to structures constructed on expansive clay soils occurs throughout the United States each year. The extent to which expansive clay soils can damage an individual foundation will depend largely on the amount and location of moisture change in the soil beneath that foundation as well as the foundation's stiffness. Significant changes in moisture levels typically occur from poor drainage and/or from large vegetation causing heave or shrinkage respectively. Once a structure has experienced damaging differential movements and repair is needed, an engineer must consider sources and magnitudes of previous and future movements in designing repairs.

This paper presents a tested technique, which involves using an aqueous solution of potassium and ammonium ions to treat expansive clay soils in remediation of damaged structures. The technique relies on reactions between the solution ions and the clay soil to adjust rheological properties of the clay. Mechanisms by which potassium and ammonium treatment reduce swelling motions and pressures are described herein. A case history of a forty-five year old building damaged by both heave and shrinkage movements is presented. Identification of movement sources, reactivity laboratory testing of site soils with the aqueous solution, predicted injection and long-term movements, description of the injection modification technique, verification quality assurance testing and foundation performance are presented.

[1] Vice President, Hayward Baker, Inc., 2510 Decatur Ave. Ft. Worth, Texas 76106; adpengelly@haywardbaker.com

[2] Consulting Geotechnical Engineer, P.O. Box 173908, Arlington, Texas 76003-3908; mbaddison@home.com

Introduction

Structures experiencing movement from expansive clays are a fairly common occurrence in many regions of the United States (Snethen, et.al., 1975). Heaving movements typically occur during wet periods that follow months of drier weather or as a result of poor drainage.

Shrinkage movements typically occur during prolonged dry periods that follow months of wet weather. Climatic moisture withdrawal from supporting soils is frequently exacerbated by root systems of nearby large vegetation. The typically elevated soil moisture conditions found under floor slabs-on-grade provide a source for tree sustenance during extended dry periods. A perennial source of nourishment near large vegetation will result in higher root density for its sustenance during low precipitation periods (Cutler, 1995).

The concentration of roots will remove more moisture from under the slab during an extended dry and active growth period for evapotranspiration than would otherwise be lost due to normal evaporation. Dissimilar soil moisture withdrawal from tree root zones in volumetrically sensitive clays frequently results in differential settlements due to shrinkage (Ravina, 1984).

After a building has experienced damaging movements, a forensic engineering investigation is typically performed to identify the cause(s) and to assist with design of repairs. Long term repairs should address the cause(s) of damage, typically moisture change in volumetrically active supporting soils, rather than temporarily releveling the affected floor area or structure etc. Drainage irregularities are relatively easy to identify and improvements are usually straightforward. On the contrary, shrinkage settlements are more difficult to diagnose and remediate, particularly when vegetation sources are suspected. In addition, remediation of tree effects can be somewhat more complicated to address. Owners may not want to remove vegetation for aesthetic, solar shading or municipal code reasons. However, it is usually not possible to adequately remediate a condition where trees are causing movement unless the vegetation is addressed. Typically, the vegetation is removed completely or the roots are severed between the tree trunk and the slab. Additionally, when the roots are severed or the tree removed, the disparate moisture regime under the slab will tend to balance with time (Holtz, 1983). The desiccated soil moisture region will slowly draw equilibration moisture from adjacent wetter regions through soil suction forces. These rehydration events will result in continual heaving motions of the settled area until some measure of equilibrium is achieved. After the soil moisture reaches this quasi-equilibrium condition, which could take years to occur, it can be maintained through proper drainage and root barriers (Biddle, 1998). Typically the owner makes frequent minor (cosmetic) repairs to the building as the destructive tree affects gradually progress. As clay rehydration occurs, these repairs are likely to require additional attention or replacement.

Minimizing clay rehydration, in specific desiccated regions, is difficult to achieve because of associated high suction forces and the natural tendency to reach equilibrium with surrounding conditions. In order to achieve a static condition more quickly, with much less rebound damage, it may be desirable to use a modification technique such as applying an aqueous solution of potassium and ammonium to the soil mass. The modification technique will also improve long-term performance of the repaired slab by making the static condition easier to maintain. Applying the modifier to the soil accomplishes this by reducing the effects that moisture changes can have on the soil's potential for volume change.

Clay Mineralogy and the Mechanism of Heave

All clay particles or minerals are composed of sheets of silica and alumina. The order in which these sheets are stacked and the ions present will determine the clay particle's mineral type and behavior. The type of clay mineral associated with heave is smectite (or montmorillonite). An example of the clay mineral smectite is shown in Fig. 1. As the parent material weathers into clay minerals, ions such as calcium, magnesium or sodium will be present in the interlayer spacing of the clay particles. These positive cations are attracted to the surface of the clay particle in an attempt to balance the net negative charge of the clay particle. Once present, the cations control the behavior of the clay particle. Swelling in clay is directly related to cation hydration energy (its attraction to water molecules) and the hydrated radius of the interlayer cations. If the hydration energy of the predominant cation is greater than the potential energy of the clay particle, the attached cation will hydrate. Hydration causes the diffuse double layer of water attached to the clay particle to grow and repel opposing diffuse layers on adjacent particles. These repulsive forces cause swelling pressure (from Norrish).

In Fig. 2-a, a model (Guven, 1992) of the surface of the clay particle is presented which shows the negative surface of the clay particle along with attached cations and the diffuse double layer. Note that the potential energy of the clay particle with distance from the surface is plotted in Fig. 2-b. It should be noted that when a predominant cation is present that has a low hydration and small hydrated radius, it will move closely to the surface and result in a rapid drop in potential energy of the clay resulting in little swell pressure (Mitchell, 1993). Alternatively, if the predominant cation has a large hydration energy or large hydrated radius (sodium, calcium and magnesium), these cations cannot get in close proximity to the surface leaving the surface potential energy high, which attracts more balancing hydrated cations resulting in increasing double layer thicknesses and increased swell pressure.

Four cations commonly occurring in soil environments, which are identified as having low hydration energies and low hydrated radii are potassium, ammonium, rubidium and cesium (Grim, 1968 and Norish, 1954). Potassium and ammonium were used as cations, which when mixed in a solution of water, can be used to modify clays beneath existing structures that are experiencing damaging volumetric movements. Test results taken on soil samples, both before and after treatment, reveal behavior consistent with theory and information described above. Clays treated with potassium and ammonium have consistently been

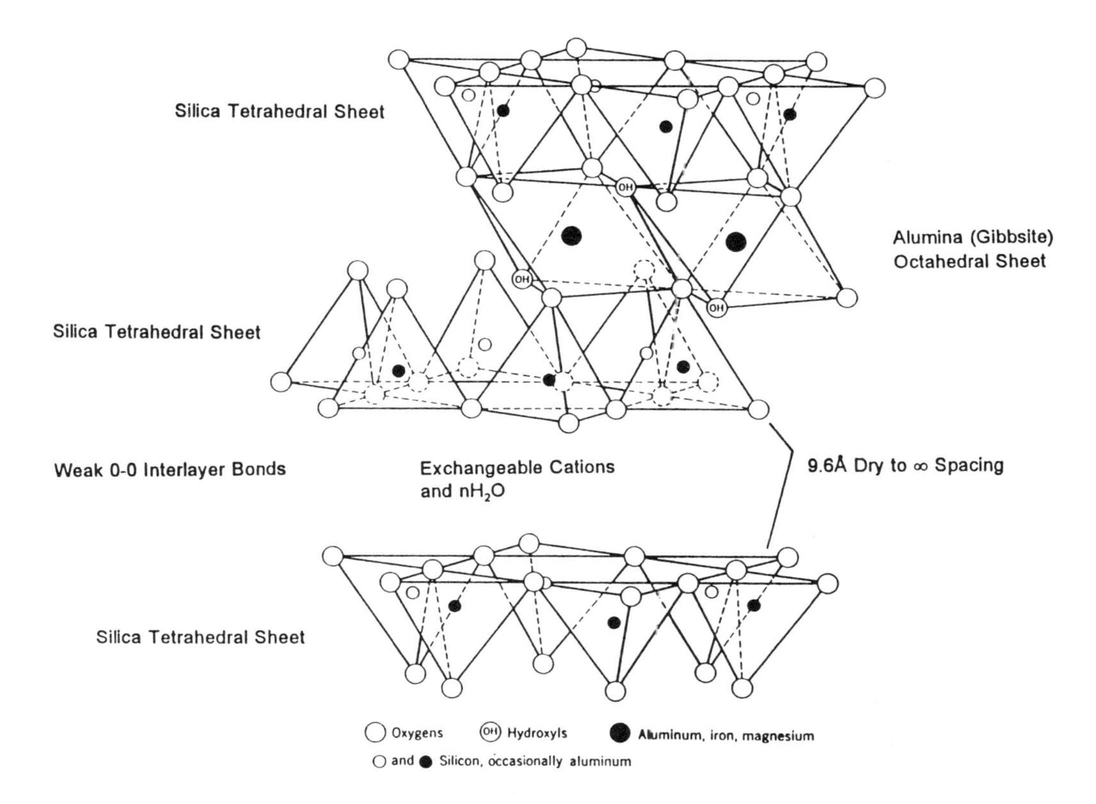

Figure 1. Structure of Montmorillonite (after Grim, 1968)

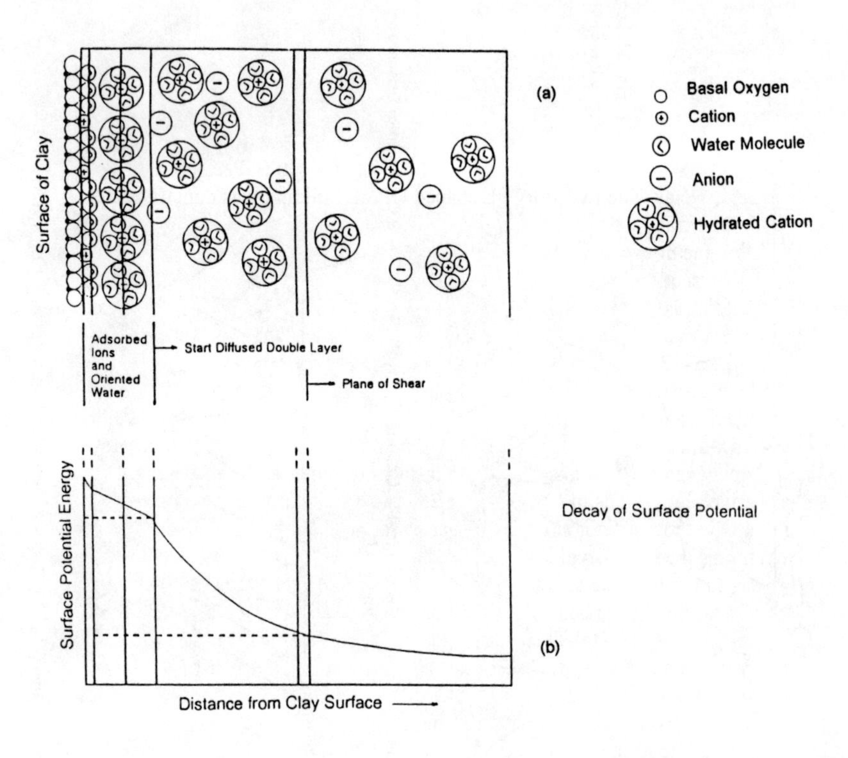

Figure 2. Multilayer Model in an Aqueous Solution (after Güven, 1992)

shown to stop swelling at lower moisture contents than untreated soil. Additionally, swell caused by the introduction of an aqueous solution containing potassium and ammonium is consistently lower than that containing water alone. This data is presented in the section titled "Reactivity Testing".

GP&L Building

One such case history where an active clay soil was modified beneath an existing structure is that of the Garland Power and Light (GP&L) building in Garland, Texas. The two story , L shaped GP&L building was constructed in 1956. The frame is supported by perimeter grade beams on drilled piers founded in limestone near a depth of ten feet. The concrete floor slab is reinforced by thickened sections, which were intended to act like grade beams. A plan view of the building is shown in Fig. 3. Along the east side of the building was located a row of six mature live oaks with one mature live oak on the north side. The 305 to 457 mm (12 to 18 in.) diameter tree trunks were approximately 2.4 m (8 ft.) from the east wall of the structure. On the west side of the east wing was a poorly drained alley way. Inside the building, distress was evident, particularly along the eastern wall. Cracks in the dry wall were evident, as were warped door jambs and voids between the bottoms of walls and floors throughout the east wing. During extensive renovations in the 1980's, a 25 to 76 mm (1 in to 3 in.) leveling course was applied to the floor slab. The floor slab continued to exhibit settlement and was mudjacked to relevel it in 1987 and again in 1991. It was the intent of the owners to remodel the space again. It was determined that the condition causing the movement should be remedied before planned remodeling took place.

Soil borings were taken along the east wall and the southern portion of the west wing and a test pit was excavated along the west side of the east wing. The test pit and borings indicated that some stiff and dense fill was present in the upper 0.3 to 0.6 m (12 to 24 in.), which contained clay with sand and brick pieces. Numerous tree roots were also present in the test pit excavation, which was on the opposite side of the building and some 13.7 m (45 ft.) from the nearest tree. Below the shallow fill was a very stiff, fat clay to a depth of approximately 2.1 to 2.4 m (7 to 8 ft.). Beneath the clay was weathered limestone. Two representative logs of the borings are presented in Fig.4 and 5. The borings taken along the eastern wall, near the trees indicated that the moisture content of the soil was 4 to 7 points below the plastic limit. Results of swell tests run on that material ranged from 3.6 to 4.7% in free swell tests with overburden pressure. Additionally, a pressure swell test was conducted (Fig. 6) which confirmed the results of the free swell tests. The boring taken on the west side of the building indicated a similar soil strata; however, the moisture content was four points above the plastic limit. It was apparent from the data that the clay soil on the western wing of the building was in a moist condition, which was attributed to the drainage condition on that side of the building. The borings were converted to ground water monitoring wells and it was discovered that water levels rose with significant rains.

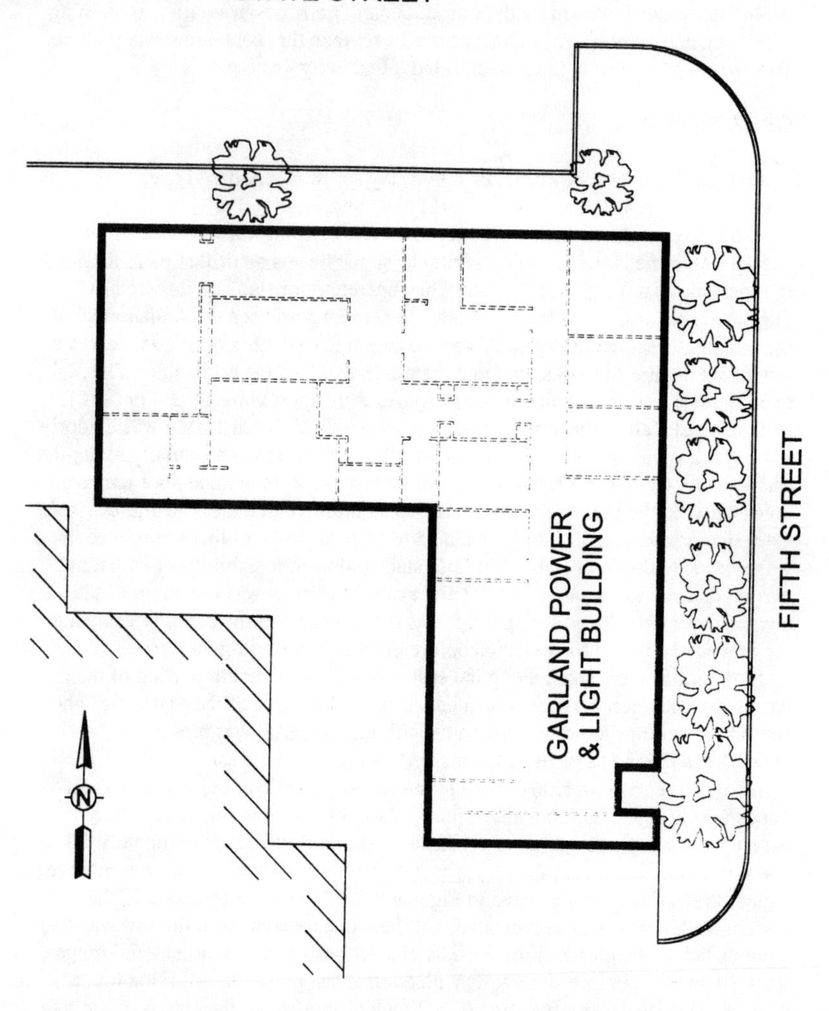

Figure 3. Plan View of GP & L Building

PROJECT: Garland Power & Light Building **Log of Boring B-1**

LOCATION: 504 State Street
Garland, Texas

Depth (ft.)	Pocket Pen (tsf)	Atterberg Limits LL / PL / PI	Moisture Content (%)	Dry Weight (pcf)	USCS	DESCRIPTION
0.00						
						CONCRETE, 4.9 in., w/ thin layer of
0.50						mud jacking materials 0.4 ft.
1.00			3.9			FILL, silty sand, tan 1.0 ft.
1.50	4.5+	77 / 28 /49	21.1	98.0	CH	
2.00						CLAY, dark brown, w/ calcareous nodules
2.50	4.5+		24.3	100.0	CH	
3.00						
3.50	4.5+		23.4	98.0	CH	
4.00						4.0 ft.
4.50	4.5+		22.7	100.0	CH	
5.00						CLAY, brown w/ calcareous nodules
5.50	4.5+		25.2	96.0	CH	
6.00						6.0 ft.
6.50	4.5+		27.5	93.0	CH	
						SILTY SAND, tan w/ limestone fragements
7.00						- refusal @ 7.0 ft. on limestone 7.0 ft.
7.50						

Completion Depth: 7.0 ft.	Project Number:	Notes:
Date: 11/20/97	Boring Method: Continuous Shelby Tube	

Figure 4. Boring Log B-1

PROJECT: Garland Power & Light Building **Log of Boring B-2**
LOCATION: 504 State Street
Garland, Texas

Depth (ft.)	Pocket Pen (tsf)	Atterberg Limits LL / PL / PI	Moisture Content (%)	Dry Weight (pcf)	USCS	DESCRIPTION
0.00						
						CONCRETE, 4.8 in.
0.50						0.4 ft.
1.00			9.3			FILL, silty sand, tan 1.0 ft.
1.50	4.5+		28.2	86.0	CH	CLAY, dark brown w/ calcareous nodules
2.00						2.0 ft.
2.50	4.5+	73 / 27 / 46	31.3	83.0	CH	
3.00						
3.50	4.5+		31.3	85.0	CH	CLAY, brown w/ calcareous nodules
4.00						
4.50	4.5+		31.9	87.0	CH	
5.00						
5.50	4.5+		31.6	87.0	CH	
6.00						6.0 ft.
6.50						
7.00						
7.50						

Completion Depth: 6.0 ft.	Project Number:	Notes:
Date: 11/20/97	Boring Method: Continuous Shelby Tube	

Figure 5. Boring Log B-2

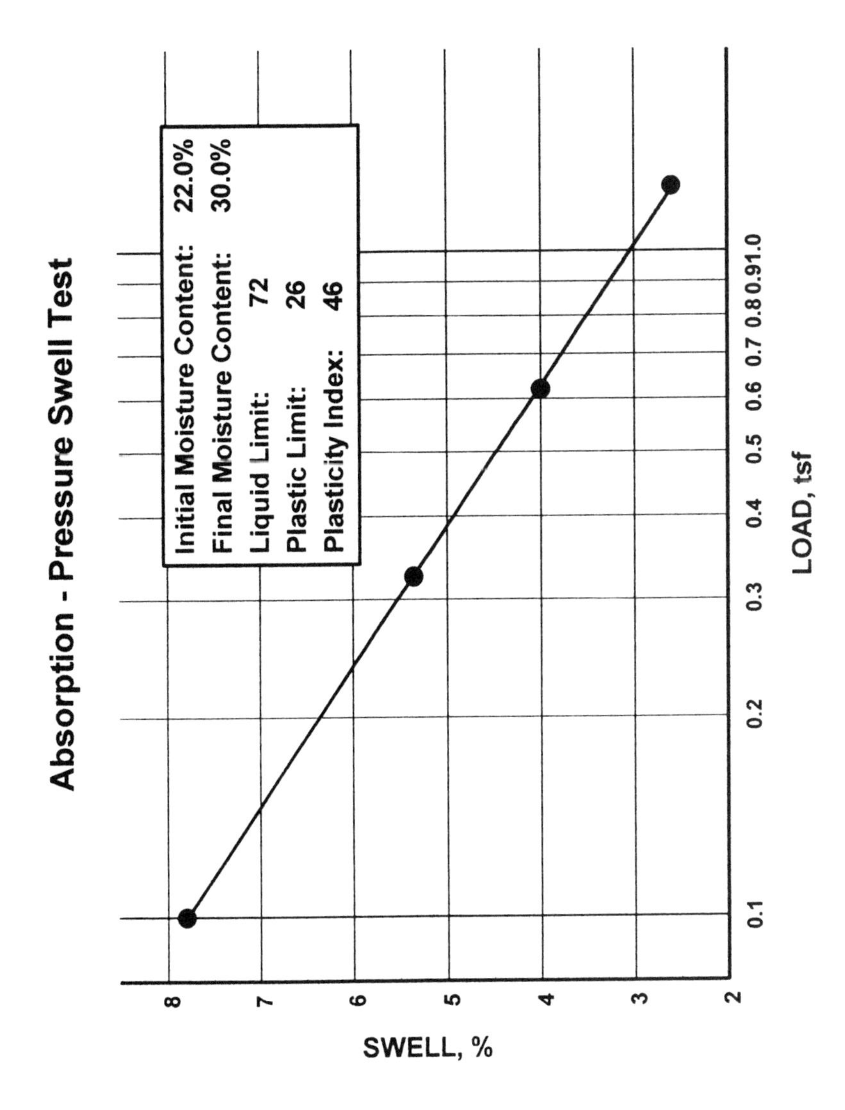

Figure 6. Pressure Swell Test

Monitoring determined that water from the poor drainage area along the alley area was traveling horizontally and vertically under the building (down gradient).

In order to remediate the settled and heaved floor conditions, it was determined that a combination of drainage and vegetation measures would need to be implemented. The roots of the live oak trees along the eastern and northern sides of the building would be severed between the tree trunks and the building and a root barrier installed. Because of the dry condition of the soil along the eastern wall, it was believed that once the roots were severed, the soil would slowly rehydrate resulting in significant damaging movements over a long period of time. If installed alone, the root barrier would also dam up subsurface seepage water under the building, inducing unacceptable heave of the dry clays. It was decided to install an interceptor drainage system along the western side of the building to intercept subsurface seepage water. In addition, the surface of that area was to be paved to minimize surface water infiltration from adjacent roof water runoff. The owners intended to remodel and then re-occupy the space with as little time vacant as possible. In order to accomplish this it became necessary to consider the use of a remediation technique that would partially re-hydrate the soil in a quick and predictable manner and would also permanently modify the clays to provide long-term stability. It was also required that the soil treatment not overly swell the soils thus nullifying the previous floor levelings from mudjacking and use of a leveling course. It was decided that a solution containing potassium and ammonium should be tested for reaction with the clay as well as to determine how much movement might be experienced by the structure during injection/absorption of the material. From the reactivity testing, the correct dosage and rate of injection could be determined for use in remediation.

Reactivity Testing

In order to perform reactivity swell tests, soil from the original borings was remolded to the same dry unit weights and moisture contents which existed in-situ. The samples were fabricated to uniform conditions into top and bottom halves designed to fit in standard consolidometer rings. Between the two halves, a layer of finely ground silica was placed and used as a simulated horizontal crack reservoir to introduce the modifying solution for absorption by the soil (Addison and Petry, 1998). Simulation of in-situ conditions as well as field injection procedures and solution quantities are essential for successfully predicting floor slab response to remediation injections. For this particular project, it was necessary to model two different in-situ conditions, most specifically the dry condition along the eastern wall and the moist condition along the western wall. While it was important to know how much swell would occur as a result of injection in the dry areas, it was also important to know if areas which were already moist would experience additional heave. Furthermore, it was necessary to determine if the material was truly reacting with the soil by comparing it to samples treated only with water. Comparisons between water treated samples and samples treated with a solution of potassium and ammonium (the concentration of solution being 25% solids by weight of the solution) were made by observing three individual measures. These were the amount of heave that occurred as a result of injecting fluid into the sample, the heave from inundation with water in a standard swell test, and finally, the post-swell moisture contents. Reactive samples will typically exhibit lower injection and inundation swells as well as a lower final moisture content.

Three sets of swell samples were created for the reactivity testing. Each set contained three samples for a total of nine tests. The first set was fabricated to model the dry condition on the eastern side of the building. The moisture contents were set at 21.9% with dry unit weights set at 1600 kg/m^3 (100 pcf). The second set of samples was fabricated to model the moist condition on the western side of the building. The moisture contents of the second group were set at 26% with dry unit weights of 1490 kg/m^3 (93 pcf). The control, or untreated set, which were to be treated with injection water (without modifying chemicals) were set at moisture contents of 21.9% and dry unit weights of 1600 kg/m^3 (100 pcf). After specimen fabrication, the samples were treated with two injections before inundation with water. Injection swell, fluid take, inundation swell and post-swell moisture were all measured. The results are presented on Table 1.

It is apparent from the results that the soil was reactive to the treatment with the potassium and ammonium solution. Evidence of modification comes in the form that nearly identical samples treated with mix water swelled 3.9 times more during absorption from injection (2.38%) compared to samples treated with the potassium solution (0.61%). Total swell, which is defined as injection swell combined with inundation swell, was 2.8 times as much for mix water specimens (4.49%) when compared to samples treated with the potassium solution (1.97%). Post-swell moisture contents for both sets of treated samples were lower than water-only treated samples by an average of 4.8%. This phenomenon is consistent in reactive soils and is also consistent with theory, which suggests that clay soils treated with potassium and ammonium will not experience excessive hydration due to the low hydration energies of potassium and ammonium.

It is also important to note from the tests that the second set of samples constructed at moisture contents of 26% actually experienced some minor consolidation during both the injection of potassium/ammonium and the inundation of the samples with water. This is attributed to the reduction in potential energy of the clay particles that occurs when potassium and ammonium are introduced into the pore water. As potassium and ammonium ions enter the inter-layer spacing of the clay particles and are exchanged with other ions, the potential energy of the particle is reduced and water is released from the diffuse double layer. This can be seen in both the consolidation as well as the fact that the post-swell moisture contents of the moist set of samples are lower after inundation with water than when they were originally fabricated.

It was concluded from results of the reactivity testing that the soil was reactive with the potassium/ammonium solution. It was determined that the heave of the

365 psf overburden pressure applied to all samples

Treatment	Sample No.	Pre-Swell Moisture (%)	Pre-Swell Dry Unit Wt (pcf)	Peak 1st Inject. Swell (%)	Total Inject. Swell (%)	Tot Fluid Take Per Sample (gal / cu ft)	Inundation Swell (%)	Total Inj + Innun. Swell (%)	Post-Swell Moisture (%)
		Initial Conditions		After 2 to 4 Days of Absorption and Injection Swelling Time			After 2 to 5 Days of Absorption and Inundation Swelling Time		
Double	1 ***	21.90	101.06	1.02	0.92	0.71	1.18	2.10	22.51
Potassium	2	21.90	99.63	0.66	0.66	0.63	1.83	2.49	23.28
& Ammonium	3	21.90	99.92	0.33	0.26	0.52	1.05	1.31	23.86
Injections	AVG. :	*21.90*	*100.20*	*0.67*	*0.61*	*0.62*	*1.35*	*1.97*	*23.22*
Double	4 ***	26.00	93.12	0.17	0.13	0.76	0.59	0.72	24.60
Potassium	5	26.01	93.56	0.39	0.26	0.63	-0.46	-0.20	25.59
& Ammonium	6	26.01	93.40	0.20	-0.07	0.72	-0.39	-0.46	25.28
Injections	AVG. :	*26.01*	*93.36*	*0.25*	*0.11*	*0.70*	*-0.09*	*0.02*	*25.16*
Double	7 ***	21.90	97.93	3.02	3.54	0.51	1.32	4.86	28.28
Mix Water	8	21.90	99.45	1.55	2.03	0.39	2.76	4.79	32.63
Injections	9	21.90	99.79	1.12	1.57	0.31	2.25	3.82	26.09
	AVG. :	*21.90*	*99.06*	*1.90*	*2.38*	*0.41*	*2.11*	*4.49*	*29.00*

***** NOTE: Fluid absorbed was adjusted between first and second injections to compensate for evaporation losses and silica layer absorption**

Table 1. Results of Pre-Injection Testing

floor in the dry areas would not be exceed 38 mm (1.5 in.). It was also determined that 6.4 mm (0.25 in) to zero heave and possibly 13 mm (0.5 in) consolidation would occur in the moist areas along the western wall. Long term stability was evident from the low swells obtained in the treated samples during inundation.

Remediation

Remediation was accomplished by three separate contractors. It was intended to install the root barrier and interceptor drains prior to performing the potassium treatment under the building. However, due to delays with the other contractors, the barrier and drain were not installed until after injection operations had been completed. It was very apparent from the excavation for the root barrier that the roots of the live oaks were well established beneath the floor slab. Roots as large as 150 mm (6 in.) in diameter were severed during installation of the root barrier. Locations of these larger roots corresponded to the lowest points of the floor slab along the eastern wall of the building. Fig. 7 provides details of the root barrier and interceptor trench that was installed around the perimeter of the building. Fig. 8 shows the locations of the root barrier and trench in plan view.

Injection of modifying solution into the soil beneath the floor slab was performed in two injection treatments. The initial injection holes were arranged at a 1.25 meter (4 ft.) center to center spacing. The holes were created by coring a 5 cm (2 in) diameter hole through the concrete slab. After the initial injection was completed, the holes were patched with non-shrink grout and a second set of holes were cored for the second injection treatment. These holes were also spaced on 1.25 meter (4 ft) centers, however, they were offset to the first set of holes resulting in an orthogonal grid. The target amount of material to be injected into the soil was set at 67 to 100 l/m^3 (0.5 to 0.75 gal/ft^3) of soil treated. The first injection was targeted to install 60% of the dosage. The second injection was to install the remaining 40%.

The injection was performed by augering a 35 mm (1 3/8 in) diameter injection rod into the soil in 6 increments of approximately 400 mm (16 in). At each downstage increment, the required amount of material was pumped into the soil and monitored at the point of injection with a flow meter. When the required amount of material was installed, the rod was advanced down to the next increment and the process repeated until the hole was completed. The rod was then withdrawn and the unit advanced to the next hole. The injection depth for treatment was 2.5 meters (8 ft) or to the top of the weathered limestone strata.

The soils were tested for swell 72 hours after the final injections were competed. Four injection verification borings were taken in the treated area. The injection specifications required four borings with three swell tests per boring. The average swell for each boring in the treated zone must be less than one percent and no one swell was to exceed 2%. This was based upon free swell tests with overburden pressures (ASTM D4546-Method B). The results were as desired with the exception of the upper most section of B-2, which contained an

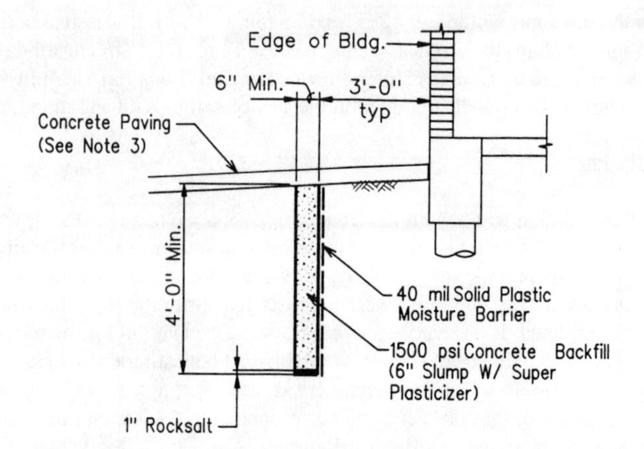

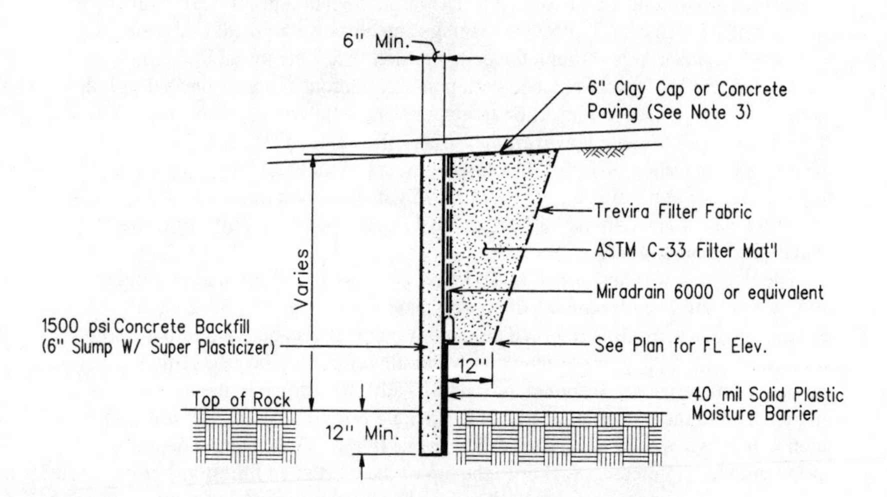

Figure 7. Root Barrier and Trench Details

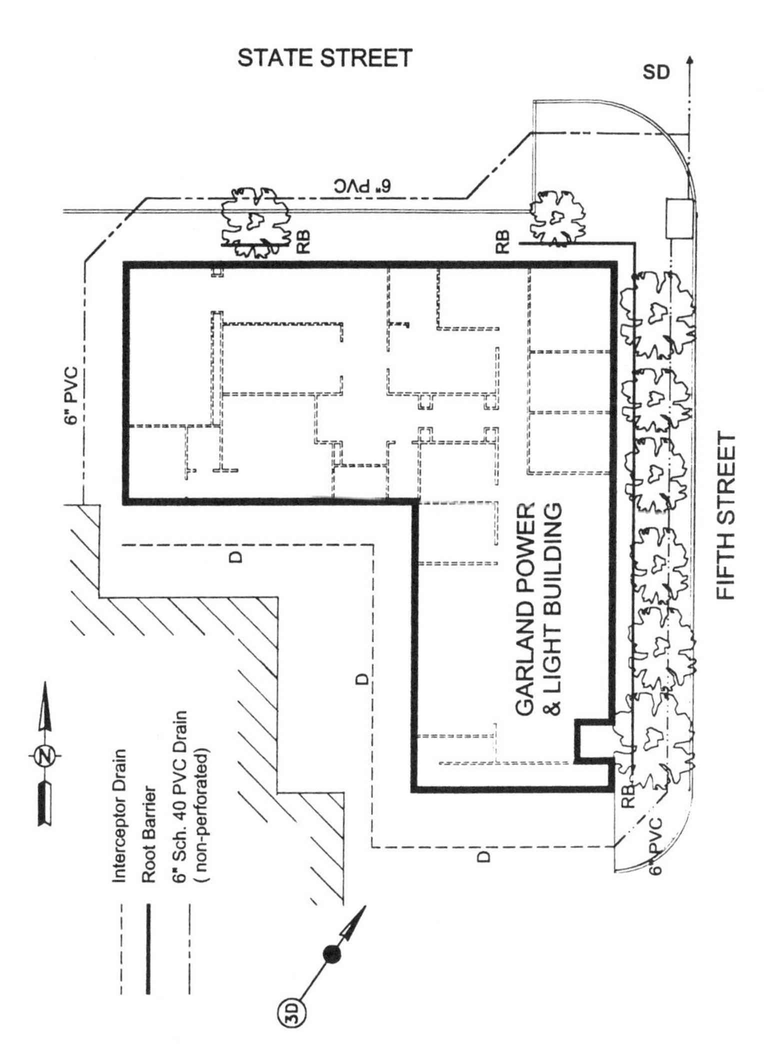

Figure 8. Plan View of GP&L Drainage Plan

individual swell of 4.67%. It was decided to retreat the entire room where B-2 was taken and retest directly adjacent to B-2 after the additional treatment. The swell in the upper section for this boring (B-5) was reduced to zero which brought all of the borings into compliance with the specification. The final average swell of all twenty tests taken was 0.43%. Results of all verification swell tests are provided in Table 2.

A total of 60,300 liters (15,932 gals) of material were used to treat an area which consisted of 1,088 m^2 (3,570 ft^2). Considering the depth of treatment was 2.4 m (8 ft), the final amount of modifying solution injected was approximately 75 l/m^3 (0.56 gal/ft^3).

Results of Treatment and Long Term Performance

In order to monitor the long term results of remediation, pre treatment, post treatment and long term relative floor elevation surveys were conducted. 3-D views of the before injection elevations are shown in Fig. 9. Fig. 10 shows the same view taken 2 weeks after the injection treatment was completed. Finally, in Fig. 11, the same view is presented again after 2.25 years. From the surveys, it was apparent that modifying injections resulted in reducing floor undulations. The floor slab was approximately one inch higher along the eastern wall after treatment. Elevations along the western wall were only slightly different. This was confirmed by viewing the cross section elevations shown in Fig. 12. Additionally, the maximum differential elevation in any one room was reduced from 38 (1.5 in) to 19 mm (0.75 in). Two and one-quarter years after treatment, another survey was conducted. Comparison of the post treatment contour plan in Fig. 10 with that taken 2.25 years later in Fig. 11 reveals substantially the same shape. Since the time the work was completed, there have been no signs of movement within the structure.

Summary and Conclusion

Expansive clay beneath shallow foundations remains a leading cause of foundation distress. This is especially true in cases where drainage and/or vegetation contribute to the problem. The paper presented a case history whereby a combination of remedial measures was used to address specific problems associated with a structure on expansive clay. These measures included drainage improvements, a root barrier and a soil treatment using an aqueous solution of potassium and ammonium. Test results both before and after treatment with the solution indicated its reactivity with the site soil as well as its effectiveness during the application. Post construction survey data was presented, including a 2.25 year survey, which indicated that the remedial measures have been effective in reducing differential elevations and limiting movement since implementation.

Boring No.	Depth Feet	In-Situ Dry Unit Weigh (pcf)	In-Situ Water Content (%)	Final Water Content (%)	Surcharge Pressure (psf)	Vertical Swell (%)	Pocket Penetrometer (tsf)
1	1	91	23.3	27.6	125	0.67	2.25
1	2	82	34.4	39.1	250	1.60	3.25
1	4	79	36.6	38.5	500	0.27	1.75
1	6	102	21.4	25.1	750	0.00	2.50
2	1	97	21.2	27.3	125	4.67	4.50
2	3	92	27.2	30.1	375	1.07	4.50
2	5	103	15.0	16.1	625	0.53	4.00
2	7	98	24.7	25.5	875	0.13	3.00
3	2	88	31.6	32.8	250	0.80	-
3	3	91	29.9	31.0	375	0.40	3.50
3	4	92	29.4	30.3	500	0.53	2.50
3	7	94	29.5	29.6	875	0.00	2.50
4	2	86	33.4	34.5	250	0.53	4.50
4	4	93	29.1	29.5	500	0.53	2.00
4	5	93	29.3	29.6	625	0.27	2.50
4	6	94	28.2	28.6	750	0.40	3.00
5	1	94	20.4	24.6	125	0.00	3.00
5	3	89	30.8	33.3	375	0.13	3.25
5	5	87	29.8	31.9	625	0.27	3.75
5	7	92	27.8	29.4	875	0.40	2.75

Table 2. Test Results of Treated Soils at GP&L

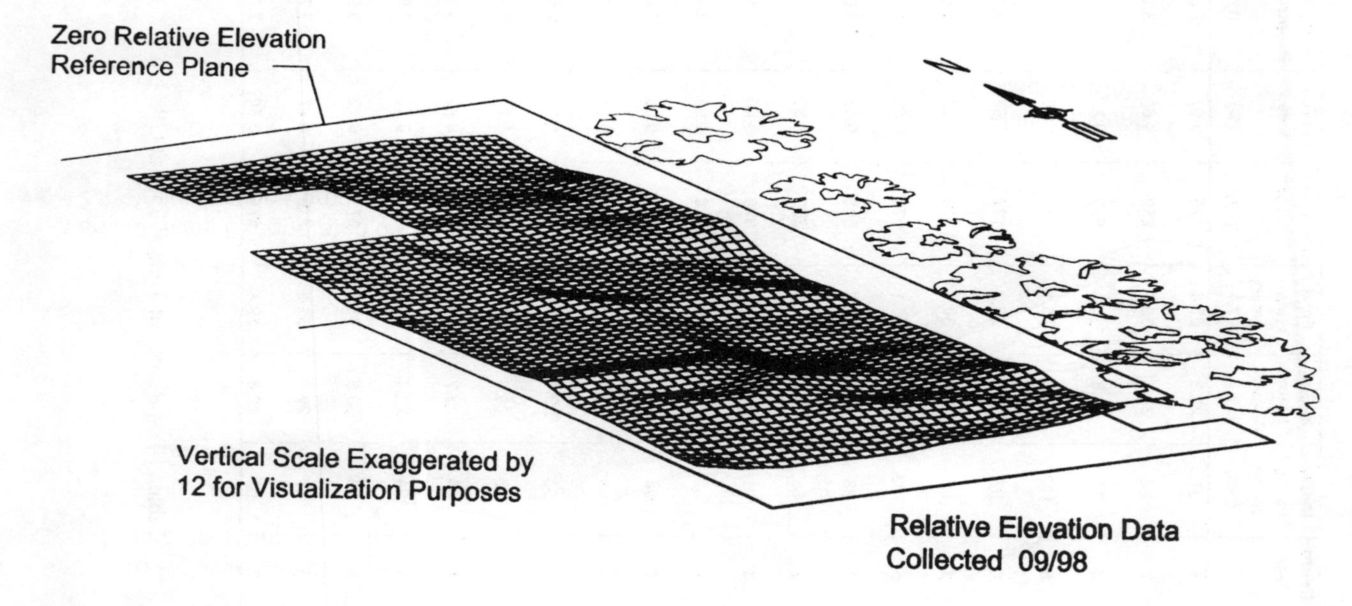

Figure 9. 3-D View of Floor Before Treatment

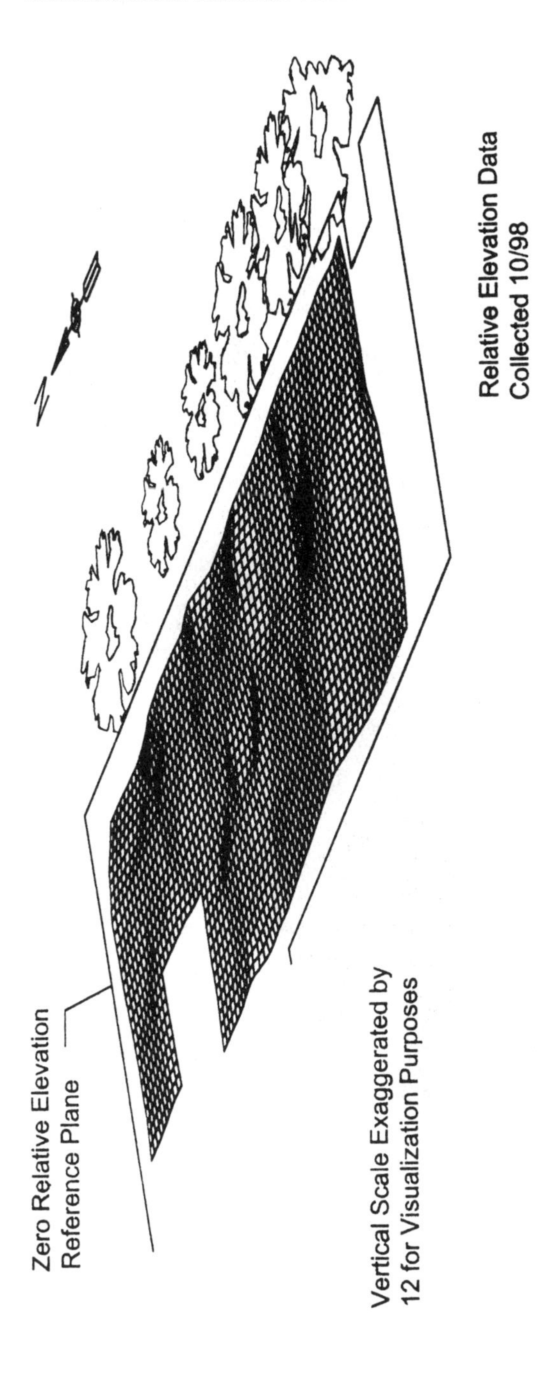

Figure 10. 3-D View of Floor Just After Treatment

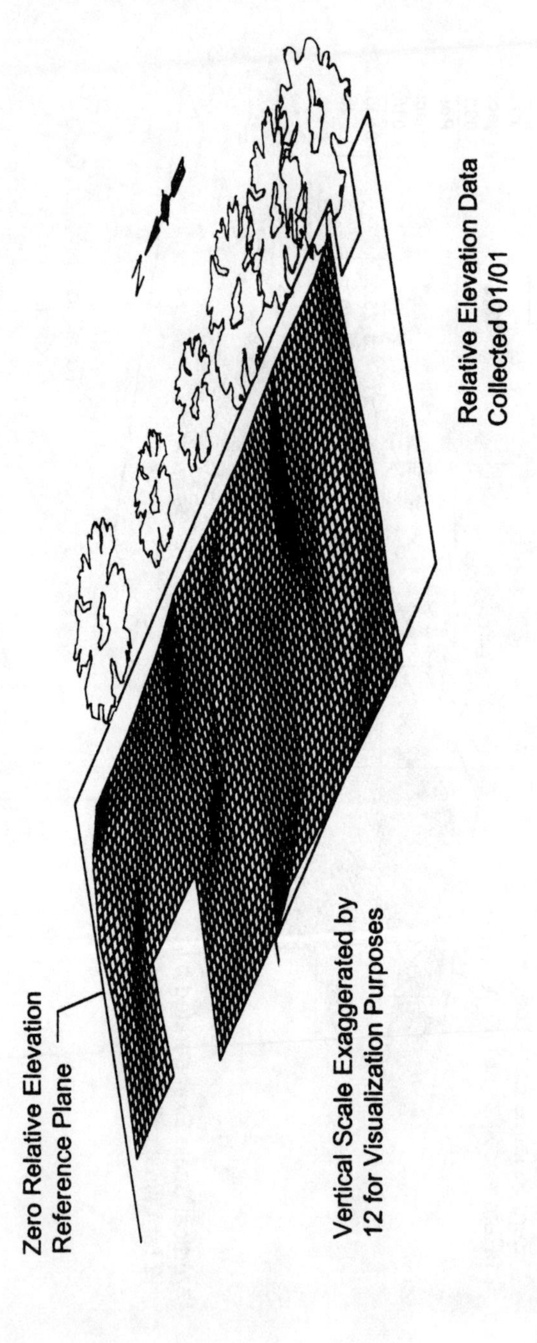

Figure 11. 3-D View of Floor 2.25 Years After Treatment

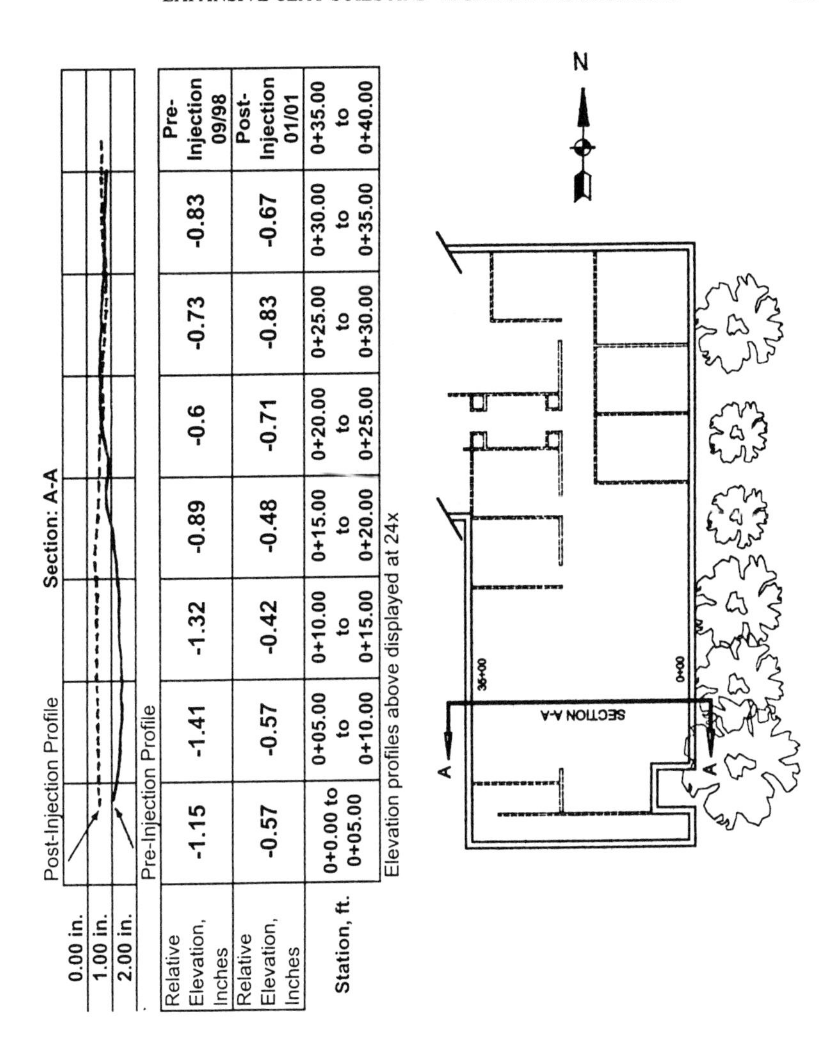

Relative Elevation, Inches	-1.15	-1.41	-1.32	-0.89	-0.6	-0.73	-0.83	Pre-Injection 09/98
Relative Elevation, Inches	-0.57	-0.57	-0.42	-0.48	-0.71	-0.83	-0.67	Post-Injection 01/01
Station, ft.	0+0.00 to 0+05.00	0+05.00 to 0+10.00	0+10.00 to 0+15.00	0+15.00 to 0+20.00	0+20.00 to 0+25.00	0+25.00 to 0+30.00	0+30.00 to 0+35.00	0+35.00 to 0+40.00

Figure 12. Cross-Section and Relative Elevation Data

Acknowledgments

The authors wish to express their thanks to GP&L for granting permission to publish the findings and to Frank A. Polma of R-Delta Engineers for providing all the general civil engineering aspects on this project.

References

Addison, M.B. and Petry, T.M., 1998. "Optimizing Multiagent, Multi-Injected Swell Modifier", Transportation Research Record, No. 1611, Stabilization and Geosynthetics, Transportation Research Board, Washington, D.C. pp 38-45.

Biddle, P.G., 1998, Tree Root Damage to Buildings, Willowmead Publishing Ltd., United Kingdom.

Cutler, David, F., 1995, "Interactions Between Tree Roots and Buildings," Trees and Building Sites, edited by Gary W. Watson and Dan Neely, International Society of Arborculture, Savoy, Illinois, pp.78-87.

Grim, R.E., 1968 Clay Mineralogy, 2nd ed., McGraw-Hill, New York.

Guven, N., 1992, "Molecular Aspects of Clay/Water Interactions," Clay-Water Interface and Its Rheological Implications, CMS Workshop Lecturs, Vol. 4, The Clay Minerals Society, pp 1-79.

Holtz, W.G., 1983, "The Influence of Vegetation on the Swelling and Shrinking of Clays in the United States of America," Geotechnique, June, pp. 159-163.
Mitchell, J.K., 1993, Fundamentals of Soil Behavior, 2nd Edition, John Wiley & Sons, Inc., New York.

Norrish, K., 1954, "The Swelling of Montmorillonite, "Transactions Faraday Society, Vol. 18, pp. 120-134.

Ravina, I., 1984, "The Influence of Vegetation on Moisture and Volume Changes," The Influence of Vegetation on Clays, The Institution of Civil Engineers, Billings and Son Ltd., Great Britain, pp. 62-68.

Snethen, D.R., Townsend, F.C., Johnson, L.D., Patrick, D.M. and Vedros, P.J. 1975, "A Review of Engineering Experiences with Expansive Soils in Highway Subgrades," Report No. FHWA-RD-75-48, Interim Report, prepared for Federal Highway Administration, Office of Research and Development, Washington, D.C.

Equipping Your Personal Drought Monitoring Tool Box

Michael J. Hayes[1]

Abstract

Monitoring drought is a task made difficult by the fact that droughts vary by intensity, duration, spatial extent, and the impacts they cause. Initially, scientists developed a series of drought indices designed to monitor drought for specific purposes. Recently, a consensus is forming that a better drought monitoring approach is one that puts together an ensemble of indicators rather than relying on one indicator alone. The appropriate analogy for this technique is the collection of the appropriate tools to place into a drought monitoring tool box. Improvements in technology, such as the Internet and GIS, have made this approach possible. This article provides an overview of the drought monitoring tools available for a tool box, and ends with some suggestions for equipping a specific tool box for monitoring droughts in regions with expansive soils.

Introduction: The Challenges of Monitoring Drought

Drought is a normal part of virtually any climate. In the United States, drought occurs somewhere within the country every year, and, on average, 10% of the U.S. is either in severe or extreme drought (data supplied by the National Climatic Data Center, NOAA). Annual drought losses in the U.S. have been estimated at $6-8 billion, more than for any other natural hazard (FEMA 1995). The 1988 drought caused losses totaling $39.2 billion (Riebsame et al. 1991).

Although drought is a natural hazard, it has several characteristics that make it different from any other natural hazard. First, it is a slow-onset, creeping phenomenon. Often officials do not realize they are experiencing a drought until it is too late and they have to react

[1]Climate Impacts Specialist, National Drought Mitigation Center, 239 L. W. Chase Hall, Lincoln, NE 68583-0749; phone (402) 472-4271; mhayes2@unl.edu

to a crisis. Drought has been described as "the dog NOT barking in the middle of the night!" Who notices that? Secondly, drought impacts are usually absent of dramatic visual destruction that frequently accompanies other natural hazards. Finally, droughts usually occur over a widespread area and affect different regions and economic sectors differently. For example, the characteristics and impacts of a drought in the semi-arid western U.S. will be completely different from those of a drought in the mid-Atlantic states or the Great Lakes region.

Each of these characteristics of drought contributes to the challenge that goes along with monitoring drought. In addition, drought severity is dependent on three components: drought intensity (the extent of precipitation deficiency), spatial extent, and duration. Therefore, monitoring drought relies on the capability to determine the extent of each of these components.

In the past, scientists have tried to find the particular measurement or index that would define and monitor drought, each with a varying degree of success. More recently, it has become clearer that, given the complexities of drought, many tools are needed to comprehensively monitor drought. This concept is similar to a person equipping a tool box with the tools necessary to accomplish a task. The idea of this paper is illustrate the importance of a drought monitoring tool box and to begin to identify the tools that engineers investigating the impact of drought on the expansive soils of a region would find necessary to include into their tool boxes to monitor drought.

Drought Indices as Tools

When scientists discovered that precipitation indicators such as total departure or percent of normal were not adequate to monitor the complexities of drought, they turned to the development of various drought indices. The goals of a particular drought index are to 1) describe the complex drought issue in a simple, relatively conceivable way; 2) provide a measurement of the drought severity components of intensity, duration, and spatial extent; and 3) put a current drought event into a realistic context with historical conditions.

Hayes (1998) provided a commentary on a series of drought indices that have been used in the U.S. over the past 35 years, along with their strengths and weaknesses. In the context of this particular paper, it is appropriate to talk about three of the indices most commonly used in the U.S.: the Palmer Drought Index (PDI), the Crop Moisture Index (CMI), and the Standardized Precipitation Index (SPI).

A. Palmer Drought Index

The PDI was developed in 1965 and has since become the drought index of choice in the U.S. (Palmer 1965). As Palmer conceived his index, he correctly understood that drought in a particular location results from a complex combination of precipitation, temperature, and soil components. His index, then, is based on the supply and demand concept of the water balance equation that incorporates Thornthwaite's approximation of evapotranspiration (ET) and available water content of the soil. The PDI does a relatively good job of accomplishing the goals necessary for an index to be successful. The index's simple scale of above and below zero values is recognized by scientists, policy makers, the media, and the public. These values were standardized (in theory) in place and time so that a -4.0 PDI value represents the same relative drought impact at any location or for any season, allowing a measurement of a drought's intensity, duration, and spatial extent.

In addition, a good long-term data set of PDI values at a Climate Division level has been established going back to 1895, providing an opportunity to place drought events into historical perspective. This data set is available on the National Climatic Data Center's website at http://www.ncdc.noaa.gov/onlineprod/drought/main.html. Weekly PDI values for the Climate Divisions in the U.S. are available on the World Wide Web at http://www.cpc.ncep.noaa.gov/products/analysis_monitoring/regional_monitoring/palmer.gif.

Unfortunately, the PDI has some major weaknesses that limit its ability as a drought index. These weaknesses illustrate why it should be used with caution and as only one tool within a drought monitoring tool box. The limitations that most affect the engineering implications of expansive soils are the following:

- *Complexity*. The PDI is a very complex index, requiring more than 60 calculations in order to determine a PDI value for a single location. This complexity limits the application of the PDI for local and specific issues such as a location with expansive soils.
- *An inherent time scale*. Although it is not known exactly what time period a PDI value represents, which is a weakness in itself, there is an inherent time scale of 8-12 months built into the PDI values. What this means is that the PDI is slow to determine developing drought conditions in a region, and if decisions are being based on the PDI values, these decisions could be very untimely and made well after a drought has begun.
- *An inaccurate representation of the runoff component*. For an application that depends on soil characteristics, and with an index that calculates runoff, it is important that this is done well. Unfortunately, the PDI operates more like a sponge than what really happens: no runoff is allowed until the sponge (soil) is saturated. In applications where decisions are made based on soil properties, this would have a considerable impact.
- *Aggregation of soil characteristics*. The PDI generalizes all soil characteristics into one type representative of an entire Climate Division in order to calculate a value. Again, given the variability of soils across any region, decisions that involve soil properties will not have the appropriate detail that would likely be needed.

In addition, there is another important weakness that, although perhaps not applicable to Texas, does apply to some of the expansive soils in Colorado, Wyoming, and other locations. The PDI does not account for frozen ground or for snowfall in its calculations, which may completely change the dynamics of the soil moisture situation within these soils.

B. Crop Moisture Index

After developing the PDI, Palmer saw a need to develop an index that would respond more rapidly to existing conditions for agricultural applications, so he developed the CMI (Palmer 1968). The CMI is actually derived from the PDI and is calculated weekly. As an index, it is more responsive to the weekly changes of precipitation and temperature. One can see the current CMI maps for the U.S. Climate Divisions online at http://www.cpc.ncep.noaa.gov/products/analysis_monitoring/regional_monitoring/cmi.gif.

Because of its relationship to the PDI, the CMI carries along with it some of the PDI limitations. In addition, the index is meant to be applied during the growing season. It was not designed to be calculated during the winter months and is reset to 0.0 to start each season. This loses any of the preseason information that may have an important contribution to the soil moisture situation.

C. Standardized Precipitation Index

In order to address some of the limitations characteristic of the PDI, McKee et al. (1993, 1995) developed the SPI. Compared to the PDI, the SPI is a relatively simple index to compute. It is based solely on precipitation and, like the PDI and CMI, its scale varies around zero, representing precipitation anomalies that are standard deviations from the median. Because it is based on the normal distribution of a long-term data set, the SPI values represent specific probabilities that provide the frequency occurrence of a particular drought event (Table 1).

Table 1. Drought Event Frequency Based on SPI Values.

SPI Values	Category	Number of Times in 100 Years	Severity of Event
0.00 to -0.99	Mild Dryness	~33	1 in 3 years
-1.00 to -1.49	Moderate Dryness	~10	1 in 10 years
-1.50 to -1.99	Severe Dryness	~5	1 in 20 years
≤-2.00	Extreme Dryness	2-3	1 in 50 years

The SPI's greatest strength is that it has temporal flexibility. An SPI value can be calculated for any time scale from a 1-month period out to longer time scales such as 12 or 24 months. This capability allows the SPI to measure precipitation anomalies that on a shorter scale (such as 1 or 3 months) might represent agricultural applications, or on a longer scale (such as 6, 12, or 24 months) might represent hydrological supplies indicated by streamflow, reservoir, or groundwater levels. This multiple time scale flexibility allows for the early detection of drought, as well as the opportunity to monitor the complex precipitation dynamics occurring within a drought (Hayes et al. 1999).

Monthly maps of the SPI are available for a selection of time scales on the NDMC website [http://enso.unl.edu/ndmc/watch/spicurnt.htm] and for many time scales on the Western Regional Climate Center website [http://www.wrcc.sage.dri.edu/spi/spi.html]. Currently, efforts are under way to improve the timeliness of the SPI's availability and the spatial resolution of the index so that it can be interpolated on a station-by-station basis instead of at the Climate Division level.

Because it is based only on precipitation (no allowance for runoff or evapotranspiration, etc.), the SPI should also be looked at as only one tool that can be placed into a person's drought monitoring tool box.

The Drought Monitor: A Tool Box Application

The Drought Monitor (DM) product is designed to give a weekly assessment of current drought conditions for the continental U.S., Alaska, Hawaii, Puerto Rico, and the Pacific Islands, and is perhaps the best example of a drought monitoring tool box. Its production is a joint effort by authors from four organizations: the U.S. Department of Agriculture (USDA), the Climate Prediction Center (CPC), the National Climatic Data Center (NCDC), and the National Drought Mitigation Center (NDMC). It was first released in August 1999 and has captured a large amount of attention since then.

Production of the DM is based on a mix of objective and subjective information. Both types of information are critical for the tool box. The objective tools used in creating the DM include the PDI, the CMI, a soil moisture model used by the CPC, streamflow information collected by the U.S. Geological Survey, percent of normal precipitation, USDA topsoil moisture information, and a satellite vegetation index. One of the major subjective tools utilized within the DM is a network of more than 100 "local" experts from states and agencies across all regions who add their knowledge about the drought situations in their areas. This subjective information is extremely valuable. Additional information includes snowpack levels in the western U.S., reservoir levels, the SPI, range conditions, and groundwater levels.

All of this information, or "tools" in this case, go into a tool box to help the DM authors provide a simple assessment of the current drought conditions that can be understood by policy makers, the media, and the public. A four-level scale was established to show the severity of drought conditions: D1 (moderate), D2 (severe), D3 (extreme), and D4 (exceptional). A fifth category, D0 (abnormally dry), was established as an alert that a region was heading into a potential drought, or that a region was recovering from drought but possibly experiencing some lingering impacts. Current DM maps are available at http://enso.unl.edu/monitor/monitor.html, and a more extensive website showing the information tools that go into the creation of the DM maps, as well as forecast information and an archive of DM maps back to May 1999, is located at http://enso.unl.edu/monitor/. This website is currently receiving up to 30,000 hits per week.

Additional Tool Boxes for Drought

A couple of additional examples of drought tool boxes exist. The first is not a tool box for drought monitoring, but rather is an excellent example of how a Geographic Information System (GIS) might be used to help create a drought monitoring tool box. Wilhelmi (1999) completed research examining the feasibility of using a GIS to develop a four-tooled agricultural vulnerability assessment for the state of Nebraska. In her study, four factors defined vulnerability: climate, soils, land use, and access to irrigation. The tools, or layers in the GIS, representing these factors included the probability of seasonal crop moisture deficiency for climate, soil root zone available water-holding capacity for soils, a USGS land use database, and a digitized layer of irrigated area for the state. These layers were combined in the GIS to create a four-category characterization of agricultural vulnerability at a 200-meter resolution. Wilhelmi called these four categories "low," "low-to-moderate," "moderate," and "high."

Research is continuing at the University of Nebraska that builds on the concepts of Wilhelmi's work and uses a GIS for a drought monitoring tool box. Wu (2001) is developing a GIS tool box for identifying the potential drought risk to farmers in Nebraska before the growing season and as the growing season progresses. Tools included in this research include the SPI; a satellite vegetation index; and soil, agronomic, and irrigation information. Again, the reasoning behind this research is that one tool alone will not be able to identify the complexities of drought, but a GIS is a great instrument to serve as a tool box and collect these tools so that they can be used together to provide important information to decision makers.

Finally, there are a couple of products that help with the creation of the weekly DM maps that act, by themselves, as drought monitoring tool boxes. Both of these products are produced by the CPC and are called Objective Blend maps. The concept behind these maps is to take several individual products that have the same scale (in these cases it is the Climate Division level) and combine them to produce a composite map with these tools blended together. The first Objective Blend map combines the PDI percentiles, CPC soil model percentiles, and 30-day precipitation into one map. The second Objective Blend map uses the first three tools, but also adds a satellite vegetation index as a fourth tool. These two maps then serve as additional tools for creating the DM maps. Unfortunately, neither of the Objective Blend maps is available to the public at this time.

Equipping a Tool Box for Expansive Soils

It can be said, with almost complete certainty, that no single indicator for monitoring drought will be able to adequately determine the drought conditions at a scale necessary to give engineers insight into the impacts that will occur at building sites in areas with expansive soils. Therefore, the suggestion is to develop a drought monitoring tool box that is specifically designed for this purpose. What would such a tool box look like? There are several key issues to consider.

First, the data needs, as well as the real-time and historical availability of these data, for this tool box must be determined. Some of these data may exist already, while others may need to be created. For example, precipitation and temperature data are going to be especially critical, preferably on a fairly small scale. In addition, soil moisture levels are probably critical for this particular application. Some soil moisture monitoring may already exist. However, this is likely to be far from adequate. Most soil moisture sites are established for agricultural uses, and therefore cover the top 100 cm of soil. For engineers placing a foundation in an area with expansive soils, soil moisture information is needed down to 5 m (16.5 feet), if not 10 m (33 feet)!

A second consideration would be whether or not a GIS would be appropriate for building this tool box. For example, a detailed soil properties layer may exist at a scale beneficial for engineers that could easily be placed into a GIS and become the foundation of a tool box created for monitoring droughts. The Natural Resources Conservation Service (NRCS) is in the process of creating detailed soil properties data for GIS applications for all counties in the U.S. (NRCS 1999). Currently about half of the counties in Texas (including most in the Houston area) have been included in this Soil Survey Geographic (SSURGO) data base. This level of data has been specifically designed for the landowner, municipalities, and natural resource planners. Precipitation, temperature, and soil moisture information can then be placed into the GIS as additional layers to help develop the tool box. Another possible consideration would be to identify any models that may exist that would help provide key information for the tool box. These models could be specific enough to look at the added effects of vegetation such as large trees, for example.

A key consideration would be to identify the resources available to help develop this tool box. This would include personnel and financial resources. Funds may be needed to support university or private research involved in developing appropriate models or building the GIS. Funding may also be needed to supplement or create soil moisture and precipitation networks. Examples of these types of networks exist across the country. Oklahoma, for example, has a very elaborate automated weather station network that has placed at least one station in every county across the state. This network then provides weather and soil information at a scale unseen anywhere else in the country.

One suggestion that seems logical as part of building a drought monitoring tool box for applications relating to expansive soils is to establish a communication network of engineers, researchers, and contractors that would report soil properties and moisture values at fixed levels to a central location so that this information could become part of the tool box used by engineers and decision makers across a broad region. This concept of a communication network would likely provide the most valuable information within a tool box, and would be similar to the "local" network that is part of the DM tool box.

Conclusions

Monitoring drought is not easy. This is because drought is very complex. The definition of drought, for example, has to be location- and sector-specific. Because drought lacks one specific definition, it becomes difficult to determine if a drought is occurring, its severity, and when it is over. Therefore, there is no single indicator for monitoring drought. Instead, it is recommended that a tool box of drought indicators, or tools, be created that is dependent on the application. Thus, it is recommended that engineers and others working on drought-related impacts and issues in areas of the U.S. where expansive soils exist design a drought monitoring tool box specific to the properties of expansive soils.

Equipping such a drought monitoring tool box specific to certain applications may not be easy. But it is really the only option that is available to enable us to track the severity of drought and be prepared for the impacts drought may cause.

References

Federal Emergency Management Agency, 1995. National Mitigation Strategy: Partnerships for Building Safer Communities. Washington, D.C., 26 pp.

Hayes, M. J., 1998. Drought Indices. National Drought Mitigation Center. Lincoln, Nebraska. [http://enso.unl.edu/ndmc/enigma/indices.htm].

Hayes, M. J., M. D. Svoboda, D. A. Wilhite, and O. V. Vanyarkho, 1999. Monitoring the 1996 drought using the Standardized Precipitation Index. *Bulletin of the American Meteorological Society*, 80(3): 429-438.

McKee, T. B., N. J. Doesken, and J. Kleist, 1993. The relationship of drought frequency and duration to time scales. Preprints, *Eighth Conference on Applied Climatology*, Anaheim, CA, American Meteorological Society, 179-184.

McKee, T. B., N. J. Doesken, and J. Kleist, 1995. Drought monitoring with multiple time scales. Preprints, *Ninth Conference on Applied Climatology*, Dallas, TX, American Meteorological Society, 233-236.

Natural Resources Conservation Service, 1999. Soil Survey Geographic (SSURGO) Data Base. [http://www.ncg.nrcs.usda.gov/ssurgo.html].

Palmer, W. C., 1965. Meteorological Drought. U. S. Department of Commerce Weather Bureau Research Paper 45, 58 pp.

Palmer, W. C., 1968. Keeping track of crop moisture conditions, nationwide: The new Crop Moisture Index. *Weatherwise*, 21: 156-161.

Riebsame, W. E., S. A. Changnon, Jr., and T. R. Karl, 1991. *Drought and Natural Resources Management in the United States: Impacts and Implications of the 1987-1989 Drought*. Westview Press, Boulder, CO, 174 pp.

Wilhelmi, O. V., 1999. Methodology for Assessing Vulnerability to Agricultural Drought: A Nebraska Case Study. Ph.D. Dissertation, University of Nebraska-Lincoln, 136 pp.

Wu, H., 2001. Personal communication. February 16, 2001.

Data Base on Under-Slab Moisture Contents in San Antonio and Houston Areas, Texas

John Styron, Philip King, Gregory Stieben,
G. Alberto Arroyo, PhD., and Manuel Diaz, Ph.D.[1]

Abstract

Movements caused by expansive clay soils continue to contribute to major foundation concerns for residential homes and commercial buildings. The original observed effects of expansive clay formations were discovered a relatively short time ago, in the 1920's and 1930's. Though many studies have been done, the profession is still in its youth and more information is needed to provide improved preventative designs.

An important inquiry to be addressed for foundation investigations in expansive clay is "What is the future expected natural moisture content under the slab sometime after construction?" The amount of local precipitation and the average rate of evapotranspiration control soil moisture and groundwater levels in underdeveloped areas. A balance of precipitation and evapotranspiration will result in a quasi-static moisture and groundwater level in the soil. However, a slab-on-grade foundation interferes with this natural balance. Soils beneath a slab are normally at a somewhat higher moisture content than soils at the same depth away from the slab.

The purpose of this paper is to provide information related to the boundaries of moisture content and volumetric condition for clay formations in the San Antonio and Houston areas. This information can eventually be used to develop improved shrink/swell estimation procedures. This paper presents data from laboratory tests performed on over 10,000 samples obtained from foundation investigations related to in-situ soils beneath existing commercial and residential slab-on-grade foundations. The investigations were performed in and around the San Antonio and Houston areas.

Introduction

Movements caused by the volumetric shrink and swell of expansive clays are a major source of foundation problems for residential and commercial structures in the

---

1 Respectively, Graduate Engineer, ASCE Associate Member (jstyron@fugro.com); Regional Vice President, ASCE Member (pking@fugro.com); Manager-Geotechnical Engineering, ASCE Member (gstieben@fugro.com), Fugro South, Inc., San Antonio, TX; Professor, ASCE Member (aaarroyo@utsa.edu); and Associate Professor, ASCE Member (mdiaz@utsa.edu), University of Texas at San Antonio.

State of Texas, as well as various other regions worldwide. Enormous amounts of damage have been reported worldwide as a result of the natural periods of drying and wetting of these expansive soils (Chen, 1988). The damage can usually be accredited to an inadequate foundation design, which is due, primarily, to the use of unreliable methods of calculating the state of swell and the shrink/swell potential of expansive clays (Chen, 1992).

Currently, a reliable method for the determination of the relative state of swell of foundation soils does not exist. The primary reason for this is that lower and upper boundaries for the volumetric state of swell of soils beneath foundation slabs that are site-specific have not been established. However, a large database of laboratory test results is now available from geotechnical studies of residential foundation slabs in the San Antonio and Houston areas that can be utilized to define these boundaries. The primary purpose of this study is to attempt to define the boundaries of the upper (swell) limit and lower (shrinkage) limit for soils beneath slab foundations in both the San Antonio and Houston areas and to evaluate the state of swell of the foundation soils at any given site in relation to these limits. The database to be used for this study included laboratory test results from over 3,000 soil borings taken from the San Antonio area and a nearly equivalent number of soil borings taken from the Houston area.

If it is possible, based on the large volume of data that has been thus far collected, to establish approximate shrink and swell boundaries for local formations and climatic conditions, a technique can then be developed to estimate potential heave or shrinkage for these local sites. Ultimately, the methods developed here can be adapted to provide a universal and reliable method for designing new, lightly loaded structures to resist the stresses created by shrinking and swelling clays.

Moisture Migration Beneath Slabs

The amount of local precipitation and the average rate of evapotranspiration control soil moisture and groundwater levels in undeveloped areas. A balance of precipitation and evapotranspiration will result in a quasi-static moisture and groundwater level in the soil. During periods of wet weather, the precipitation may exceed the evapotranspiration rate, resulting in elevated soil moisture and a rising groundwater level. A similar situation may occur if extensive irrigation is done in the area. During arid periods, the soil will tend to dry out and any groundwater level will be lower. However, a slab-on-grade foundation interferes with this natural balance (Nelson etal., 1992).

The presence of the slab will greatly reduce the evapotranspiration rate, but only somewhat reduce the inflow due to precipitation or irrigation because of groundwater's ability to migrate laterally. In addition, normal capillary rise will continue beneath the slab. Therefore, soils beneath a slab are normally somewhat moister than soils at the same depth away from the slab. However, an unusually wet season may result in wetter conditions away from the slab than under the slab. With time and normal precipitation patterns, the soil moisture profile will return to its normal condition of being somewhat moister under the slab. Seasonal variations in soil moisture away from the slab will generally occur fairly quickly; however, seasonal variations in soil moisture beneath the slab will be slower.

Introduction of free water to swelling soils beneath slabs can occur as a result of lateral migration or seepage of precipitation from the outside; this can be aggravated by ponded water resulting from poor drainage around the slab. In addition, leaking utility lines can create a source of free water. Soils in close proximity to a source of free water will generally be wetter than similar soils away from the source.

Foundation Movements

In general, houses and commercial structures are relatively light structures, and are, therefore, usually supported at or near the ground surface using a slab-on-grade foundation. Because the structure and the slab are relatively light, movements in slab foundations are a relatively common occurrence, and are often accompanied by cracking.

In an expansive clay area, the reason for foundation movement is that the slab-on-grade foundation produces a pressure much less than the swelling pressure of the soil, which allows for soil heave. Swell pressures of up to 1000 kPa (21 ksf) have been measured. An embankment 45 to 50 m. (150 to 165 ft) thick would be needed to resist this pressure! Pressures such as this may be out of the ordinary, but pressures of 100 to 200 kPa (2,000 to 4,000 psf) are common and a typical building would only apply a pressure of approximately 10 kPa (200 psf) per story (Holtz etal., 1981).

Uniform soil heave would generate a minimal effect on a foundation because the entire system would move as a unit, which would limit the possibility of cracking. However, the heave movements developed by expansive soils will usually be uneven because the clays are nonhomogenous and the distribution of the water infiltration beneath the slab is not uniform. For these reasons, major differential movements and, thus, major foundation damage are possible.

In addition to the lack of uniformity in the soils themselves and in the degree of moisture infiltration, there is typically variation in moisture content between the soils beneath the slab and the soils at the perimeter of the slab. The soils beneath the center of the slab are typically more moist than the soils located near the perimeter of the slab because of the reduced rate of evapotranspiration. Also, the moisture beneath the slab can increase due to capillarity and temperature differentials. The edges of the slab are less protected and more prone to seasonal moisture changes. The higher moisture conditions in the center of the slab will typically result in a somewhat higher state of swell in the active soils. This higher state of swell typically results in a "domed" shape in the slab; i.e. the center of the slab is usually higher than the perimeter (Wray, 1995). Also, the edges cycle up and down with moisture fluctuations.

Additions to existing structures are particularly prone to movements. These movements may be aggravated by: 1) foundations supporting the addition being substantially different or of lesser quality than the original structure's foundation, 2) additions often extend out a considerable distance from the existing structure and may not be adequately tied into the existing foundation, and 3) the presence of the new slab will affect the normal moisture balance and a fair amount of time may be necessary for a new quasi-static condition to develop.

The amount of foundation movement associated with soils changing from a non-charged to a charged state depends on: the clay mineralogy, the moisture content of the soil at the time of construction, the unit weight of the soil, the thickness of the active zone, the overburden provided by the foundation slab, and the amount of free water available. It is difficult to accurately estimate potential slab movement resulting from shrinkage or swelling, but for highly-plastic soils, Terzaghi and Peck reported movements in excess of 30 cm (12 in.) are possible(Terzaghi etal., 1967), and Chen reported movements in excess of 15 cm (6 in.) are possible (Chen, 1988). Leaner clays will tend to produce less movement.

Soil moisture can also be affected by vegetation (Perkins etal., 1981)(Peck etal., 1974). Trees and shrubs draw moisture from the soil through their root systems, causing localized drier areas in their vicinity. The fast growing varieties generally create the greatest demand on soil moisture. The radius of effect of a tree is related primarily to the lateral extent of its root system, which has some relationship to tree height and the spread of its branches. In general, it may be assumed that a root system has a significant effect out at least as far as the drip line (extent of branches). If the moisture withdrawn by vegetation is not replenished by precipitation or irrigation, highly plastic clays will dry and shrink. A slab built over the affected area may lose support and settle; severe cracking can result. Examples of such movement can be seen in the "roller coaster" effect of curbs and paving along tree-lined streets and in concentric crack patterns in street pavement near large trees.

Existing Heave Prediction Methods

There are many heave estimation methods available today. The decision as to which one to choose may depend on many factors. First of all, the techniques may generally be grouped into one of two types 1) analytical methods and 2) empirical methods. Although, there is no distinct defining line between the two types. The analytical techniques involve advanced, sometimes long term, tests to investigate the heave of soils with exposure to water. The actual heave of specific samples is measured and utilized to sum the total expected heave at a site. Empirical methods typically involve the use of standard tests or otherwise basic characteristics of a soil in order to estimate theoretical heave potential for a given site. Whereas the analytical methods are probably more precise, they are also typically more time consuming and more expensive than the empirical methods. So, the decision upon a method depends on the choices applicable locally, and choices that meet the given time and financial constraints.

The current methods are helpful, however, there is much room for improvement. F.H. Chen supports this opinion is his book, *Foundations On Expansive Soils*. He states, "Although the phenomenon (of heaving clays) has been fully recognized by engineers for many years, a definitive method of measuring the swelling potential of a clay has not yet been established."(Chen, 1988) The prediction of swell is especially difficult for investigation of lightly loaded structures, such as slab-on-grade foundations, where overemphasis of the applied load has disastrous implications.

One method discussed later in this paper was developed by Chester McDowell (McDowell, 1956). McDowell devised a method for estimating moisture fluctuations beneath pavements. The database he developed was compiled from a limited number of recompacted clay samples which were recovered beneath pavement sections.

Classification

A soil classification system provides the engineer a means of defining soils. The classification methodology can be used as a tool to identify expansive soils as well as to measure the degree of their expansion potential. Prior to discussion of the individual classification systems, it is first necessary to define some of the terminology used in most of the common classification systems that enables communication between engineers and classification systems to be utilized. The most important terminology connected with expansive clays is the Atterberg limits. It is apparent at this point that the addition or subtraction of water from the void spaces in expansive soils causes movement. Exactly how much water is present is important. But, it is also crucial to compare on a scale the water content of the soil versus a standard of engineering behavior. The "Atterberg limits" provide such a scale for comparison and have come to be considered as an inherent trait of a particular soil.

Atterberg Limits. The Atterberg Limits may be used to provide a general indication of the materials' potential for volumetric change (shrink/swell) with moisture variations. The liquid limit and plastic limit are collectively termed the Atterberg limits, and are derived from laboratory tests. The liquid limit is the moisture content at which the soil changes from a plastic state to a liquid state. The plastic limit is the moisture content when the soil changes from a semi-solid state to a plastic state. The numerical difference between the liquid limit and the plastic limit is termed the plasticity index (PI).

Liquidity Index. The Liquidity Index (LI) is defined as the difference between the moisture content and the plastic limit, divided by the plasticity index. The Liquidity Index provides a measure of the moisture content relative to the Atterberg Limits of the sample. For the same soil sample, the higher the moisture content, the higher the Liquidity Index. For example, an LI of zero indicates the moisture content is equal to the Plastic Limit; an LI of +0.50 indicates the moisture content is halfway between the Plastic Limit and Liquid Limit. A negative LI indicates that the moisture content is less than the Plastic Limit.

Consistency Index. The Consistency Index also provides a measure of the moisture content relative to the Atterberg Limits of the sample. The Consistency Index (CI) is defined as the difference between the Liquid Limit and the moisture content, divided by the Plasticity Index. For the same soil sample, the higher the moisture content, the lower the Consistency Index. For example, a CI of 1.0 indicates the moisture content is equal to the Plastic Limit; a CI of 0.50 indicates the moisture content is halfway between the Plastic Limit and Liquid Limit. A CI of greater than 1.0 indicates that the moisture content is less than the Plastic Limit.

Swell and Shrinkage Limits. Swell Limit is the moisture content at which an increase in moisture content will not cause an increase in volume of the soil mass, but at which a decrease in moisture content will cause a decrease in the volume of the soil mass. Conversely, Shrinkage Limit is the moisture content at which a reduction in moisture content will not cause a decrease in volume of the soil mass, but at which an increase in moisture content will cause an increase in the volume of the soil mass.

Potential Volume Change. Common practice is to relate potential volume change to Plasticity Index and the Atterberg Limits, as shown in the following table developed by Holtz (Holtz, 1959), and Dakshanamurthy and Raman, (Dakshanamurthy etal., 1973). For example, if the Plasticity Index is greater than 35, the clay is said to be highly expansive. Shrinkage Limit and Liquid Limit are not commonly used in practice at present. However, it is generally accepted that Shrinkage Limit is a lower bound for causing movements because the clay stops shrinking when the moisture content reaches the Shrinkage Limit.

Table 1. Volume Change and Index Properties

Potential for Volume Change	Plasticity Index, PI	Shrinkage Limit, SL	Liquid Limit, LL
Low	<18	>15	20-35
Medium	15-28	10-15	35-50
High	25-41	7-12	50-70
Very High	>35	<11	>70

State of Volume Change. During investigations of foundations on expansive clays, it is also important to know the state of volumetric change in the clay at the time of the investigation. A comparison of the soil moisture content with the Atterberg limits can provide an approximate indication of whether the highly plastic clays are in a shrunk state, a swelled state, or in an intermediate state of swell. The comparison can also be used to evaluate the probable additional volumetric change which could occur in the clay. To make these determinations, it is necessary to know the upper and lower limits where the clay stops swelling and shrinking. Fig. 1, developed by Coyle and Stieben (Coyle etal., 1996), conceptually illustrates these limits in relation to the other Atterberg Limits. If the moisture content is below or near the plastic limit (liquidity index near or below zero), the soils are probably in a low state of swell. However, moisture contents significantly greater than the plastic limit (higher liquidity index) may indicate a swelled soil (Sowers etal., 1967).

Presented Data Base

During the period starting in the summer of 1992 until present, Fugro South performed well over 3,000 foundation investigations related to in-situ soils beneath existing commercial and residential slab-on-grade foundations. These studies have developed large amounts of standard laboratory test data from in-situ soils sampled beneath slab-on-grade foundations. This paper presents information developed from geotechnical studies performed in the San Antonio and Houston areas.

It is believed that this data may provide a window into the range of possible in-situ soil properties in those specific regions. That is the foundation for this attempt to compile and analyze a database of this information. It is useful to investigate what this data can tell us about state of swell since it provides a portrait of the conditions and seasonal changes of the soils in these two localities. It may then be possible to expand on this information and eventually devise a method to estimate swell more accurately for the two areas indicated here as well as elsewhere.

The research provided here continues the research performed by Tom Posey and Jean Louis Briaud (Posey etal., 1995), in which the first large regional database of soil tests was developed. Although, the database was smaller and was not analyzed in the manner presented here. In the Posey/Briaud study, equations for heave estimation were formulated but boundaries were not determined and various indices such as the LI were not examined. Another difference was that the Posey/Briaud study included soil data from San Antonio and Corpus Christi, whereas the focus of this study is concentrated on San Antonio and Houston.

This database and associated research also expands on the study presented by Dr. Harry Coyle and Gregory Stieben (Coyle etal., 1996). The Posey/Briaud database was expanded upon in that study and indices such as LI were examined. Coyle/Stieben presented data showing the relationship between moisture content and Atterberg Limits for clay soils in the Central Texas area. Using LI values derived from this data, the LI values display an overall range between about -0.20 and +0.30 for the San Antonio area. That is, the soil is probably in a non-swelled state if its LI is close to -0.20, and it is likely in a fully-swelled condition if its LI is on the order of +0.30.

However, the Coyle/Stieben data was still limited, particularly in reference to analytical tests such as shrinkage and swell testing. In addition, secondary swell was estimated as the swell tests performed for their study did not continue beyond primary swell and, therefore, did not coincide with the in-situ upper boundary. This was an admitted problem in their testing procedure that warrants improvement during this current study. The swell tests performed in that study included only the effects of primary swell, and not secondary swell. The boundaries of state of swell were examined for Corpus Christi as well as San Antonio. However, the boundaries were conceived based on visual inspection of the data and engineering judgment alone, based on limited testing, and did not include any statistical evaluation.

The purpose of the development of this database is to provide a simpler and more site-specific approach to evaluating state of swell in San Antonio and in Houston and to provide a stepping stone for research which may provide possible answers to prediction of heave in these areas and eventually others as well. Given the nonhomogenous and almost unpredictable behavior of expansive clays, it is nearly impossible to develop an absolutely exact procedure for estimating heave. However, it may be possible to provide a much simpler, cheaper, and more accurate site-specific method of estimating heave than in the past.

Characterization of Data. The database includes laboratory test results from geotechnical studies performed in the San Antonio and Houston areas. Most of the data includes Atterberg limit and in-situ moisture content test results. In addition, free swell tests were performed on selected samples as a method to evaluate the shrink/swell characteristics of the soils. The swell test enables the evaluation of the approximate swell limit of the tested specimen, which is the moisture content at which the soil stopped swelling under the given effective stress. Shrinkage limit testing was also done on selected samples. The shrinkage limit results provide an indication of the lower limit of volume change of the tested sample.

Much of San Antonio data include laboratory tests performed on clay soils of the Pecan Gap geologic formation. The Pecan Gap soils are particularly expansive as illustrated by the Casagrande liquid limit versus PI plots included as Fig. 2. Much of the Houston data included laboratory tests performed on clay soils of the Beaumont geologic formation. Casagrande plots of the Houston soils included in the database for this study are presented on Fig. 3.

The plots presented on Figs. 2 and 3 suggest the majority of the soils of these formations have liquid limits ranging from 50 to 100, with an overall range of about 20 to 130. In addition, the soils generally plot between Casagrande's A-line and U-line. Although, this may not always be the case, as the figures indicate. These results suggest the Pecan Gap soils of San Antonio and the Beaumont soils of Houston are predominately inorganic, highly plastic 'fat' clays (CH), although, some samples consist of lean clays (CL) and/or silts (SM). The results also suggest that the tests fall within the normal range of expected values and are therefore generally valid.

Moisture contents versus Liquid Limit for San Antonio and Houston are presented on Figs. 4 and 5, respectively. Fig. 4 shows the moisture content values of the San Antonio data generally increase with increasing liquid limit as expected and generally range from less than 10 percent for low PI clays to nearly 50 percent for the most expansive samples. Moisture content data for the Houston soils, as shown on Fig. 5, indicates an overall range of about 5 to 55 percent, with an increase in moisture content corresponding to an increase in PI.

For both soil groups, significant sieve analysis results were not available for inclusion in this database. However, past experience in the areas suggests that the majority of the samples would have a percentage passing the U.S. Standard No. 200 Sieve of over 50, indicating the majority of the soils in these formations are fine-grained.

Figs. 4 and 5 also show graphically the variations between the moisture content versus liquid limit data presented in this current study from the results of Coyle/Stieben's and McDowell's (McDowell 1956) moisture content limits (Note: Coyle/Stieben's limits were unavailable for comparison with Houston Data). The comparisons to McDowell made in the Coyle/Stieben study are, thus, in general agreement to the comparisons made by the results of this study.

The variances between the different studies are, more likely than not, the result of the size of statistical database. It should be noted that this study had over 10,000 data points whereas the Coyle/Stieben study had approximately 6,000 data points and McDowell's study had 164 data points. In addition, this study used numerical analysis to evaluate the shrink/swell boundaries whereas the two earlier studies use visual inspection to derive best fit boundaries lines.

The upper field moisture content limits are relatively similar among the studies. However, McDowell's lower limit is considerably higher than the results indicated by this study or the Coyle/Stieben study. The reason for this is likely because the soils investigated in the McDowell study represent remolded, artificially moistened and compacted, soils beneath pavements which were not allowed to reach a dry condition.

Scatter plots showing all of the liquid limit versus liquidity index data evaluated for this study, upon basic statistical evaluation and preliminary removal of outliers, are presented on Figs. 6 and 7 for San Antonio and Houston, respectively. Also shown are on Figs. 6 and 7 are the final LI values from the free swell and shrinkage limit tests.

The San Antonio database had a similar amount of data points in comparison to the Houston database, and the trends are very nearly the same for the two data sets. Nonetheless, the San Antonio liquidity index plot is noticeably more scattered than the Houston plot. This is true for the in-situ data as well as the shrinkage and swell tests. A noticeable decrease in the slope of curvature for both the San Antonio and Houston in-situ data occurs above liquid limits of about 40. The slope change appears to be sharper for the Houston data. Generally, there is much more data and the trends are much more pronounced for this current study in comparison to the previous databases formulated by Posey/Briaud and Coyle/Stieben.

Shrink and Swell Boundaries. Statistical analysis of the two data sets developed upper and lower boundaries of shrinkage and swelling in relationship to LI versus Liquid Limit. Figs. 8 and 9 show the statistical results for San Antonio and Houston, respectively. Also shown on the figures are the lower estimated shrinkage limit boundaries using Arthur Casagrande's equation. Casagrande developed several easy methods of utilizing the liquid limit and the plasticity index of the soil to find the shrinkage limit. One of his suggested means of measurement resulted in an equation for the shrinkage limit, as indicated in Holtz and Kovacs (Holtz etal., 1981). It should be noted that the lower boundary of swell state, depicted by Casagrande's equation, stays relatively constant as the liquid limit (LL) increases.

More shrinkage limit testing was done in the process of this current research than was done in the Coyle study, and the data was analyzed statistically. However, despite improvements over past research, the author still has several reservations against the use of the lower limit of the shrinkage tests as the lower boundary for state of swell:

1. Visible inspection of the data suggests high scatter. (see Figs. 6 and 7).
2. As Casagrande pointed out, there are inherent inaccuracies within the testing procedure itself.
3. Results of statistical analysis indicated relatively poor statistical correlations (R-values) between the lower bound of the shifted/transformed shrinkage limit LI values and the Liquid Limit, in comparison to the other tests performed for this study.
4. Regression analysis results yielded poor R^2 and RMSE values compared to the other parameters evaluated.

The boundaries for state of swell in San Antonio presented in this paper are in general agreement with those found during the Coyle/Stieben study. Their upper and lower boundaries were constant values for all liquid limits of +0.30 and –0.20, respectively. The limits of state of swell developed for this current study displayed a wider range than the Coyle/Stieben bounds. On the average, the LI values in San Antonio were determined to range from about –0.30 to about +0.40 for Liquid Limit values ranging between 40 and 120. {The average boundaries for Houston range from an average of –0.30 to +0.60. The results of both the Coyle/Stieben study and this current study verify that samples can continue swelling at moisture contents well above the plastic limit, which extends beyond Sowers' observance that swell continues to LI values of 0.10 to 0.30 (Sowers etal., 1967).

Conclusions

The results of this study have provided a valuable and relatively accurate means for the estimation of state of swell. Additional conclusions can be made as well. The differences between San Antonio and Houston do, in fact, affect the shrink/swell boundaries. Therefore, these other influences, of which climatic variations may be the most critical, must be incorporated into any heave estimation technique. Another important conclusion of this research is that the upper limits of the state of swell for San Antonio and for Houston suggest soils are able to continue swelling well beyond moisture contents equivalent to the plastic limit and beyond the limits that were previously believed to exist. In fact, soils beneath slabs have the capability to obtain the highest possible state of swell, known as the secondary swell limit. This is proven by the databases produced through this research. So this must be taken as the upper limit of swell for the purposes of design. The wide range of potential movements evaluated for this study provides greater justification than ever before into continued research into more effective ways of estimating swell.

Recommendations for Future Study

The results of the research presented here provide a starting point for continued database research into the behavior of expansive clays, prediction of state of swell, and for creating optimized procedures for estimating potential heave movements. The long term goal of this study is to further define a method for evaluating swell potential by developing a correlation/relationship with standard classification tests based on experimental data with an emphasis on lightly loaded structures and with overburden and structural load taken into account. Presented below is an itemized list of areas of possible future research:

- more standard and long term swell testing to relate swell state, including secondary swell, more effectively with heave amounts;
- more and improved procedures for evaluating the lower boundary (shrinkage limit);
- supplementary research into relationship of mineralogy and clay content to the boundaries developed here;
- additional investigation into a climatic/environmental factor which explains differences in the state of swell boundaries between soil formations; and
- further assessment of topography and drainage conditions in relation to the state of swell boundaries developed for this study.

Acknowledgement

This paper is a summary of information presented in Mr. John Styron's Master of Science Thesis (Styron etal., 2000). The remaining authors of this paper extend our appreciation to Mr. Styron for the development of the database contained in his Thesis.

References

Chen, F.H., (1988), Foundations on Expansive Soils, Developments in Geotechnical Engineering, Vol. 54, Elsevier Science Publishing Company, Inc., New York., Pg. 16. Pg. 200, 52.

Chen, Fu, (1992) "Practical Approach on Heave Prediction", 7th International Conference on Expansive Soils, Dallas, Texas.

Nelson, J.D. and Miller, D.J. (1992) Expansive Soils – Problems and Practice in Foundation and Pavement Engineering, John Wiley & Sons, Inc., NY, Pg. 15.

Holtz, H.D., Kovacs, R.D., (1981), An Introduction to Geotechnical Engineering, Prentice Hall, New Jersey, Pg. 187, 182.

So Your Home Is Built On Expansive Soils, (1995), Edited by Warren K. Wray, Geotechnical Engineering Division of the American Society of Civil Engineers, Virginia, Pgs. 15 & 16.

Terzaghi, K. and Peck, R.B., (1967) Soil Mechanics in Engineering Practice, Second Edition, John Wiley & Sons, Inc., New York, Pg. 146.

Perkins, R.L. and Elsbury, B.R., (1981) "Structural Damage due to Soil Volume Change", Reprint from Soundings, McClelland Engineers, Inc., Houston, Texas, winter edition.

Peck, R.B., Hanson, W.E., Thornburn, T.H. (1974) Foundation Engineering, Second Edition, John Wiley & Sons, New York. Pg. 343.

McDowell, C., (1956), "Interrelationship of Load, Volume Change, and Layer Thickness of Soils to the Behavior of Engineering Structures", Proceedings, Highway Research Board, Washington, D.C., Vol. 35, Pgs. 754-772.

Holtz, W.G. (1959) "Expansive Clays - Properties and Problems", Quarterly of the Colorado School of Mines, Theoretical and Practical Treatment of Expansive Soils, Golden, Colorado, Vol. 54, No. 4., pp. 89-125.

Dakshanamurthy, V. and V. Raman, (1973), "A Simple Method Identifying an Expansive Soil", Soils and Foundations, Tokyo, Vol. 13, No. 1, pp. 97-109.

Coyle, H.M., and Stieben, G.P., (1996) "Expansive Clay Shrinkage-Swell Limits", Texas Section ASCE Fall Meeting, San Antonio, Texas.

Sowers, G.F. and Kennedy, C.M., (1967) "High Volume Change Clays of the Southeastern Coastal Plain", Third Pan-American Congress of Soil Mechanics and Foundation Engineering, Caracas, Venezuela, Volume II.

Posey, T.A., Briaud, J.L., (1995), "Database for Expansive Soils In San Antonio and Corpus Christi", Texas A&M University, College Station, Texas.

Coyle & Stieben.

McDowell, C.

Holtz & Kovacs, Pg. 182.

Sowers & Kennedy.

Geotechnical Research Study-Shrinkage/Swell Limits and Relative State of Swell, Expansive Clay Soils: San Antonio/Houston, Texas, Styron, John P., Thesis, The University of Texas at San Antonio, Dec 2000.

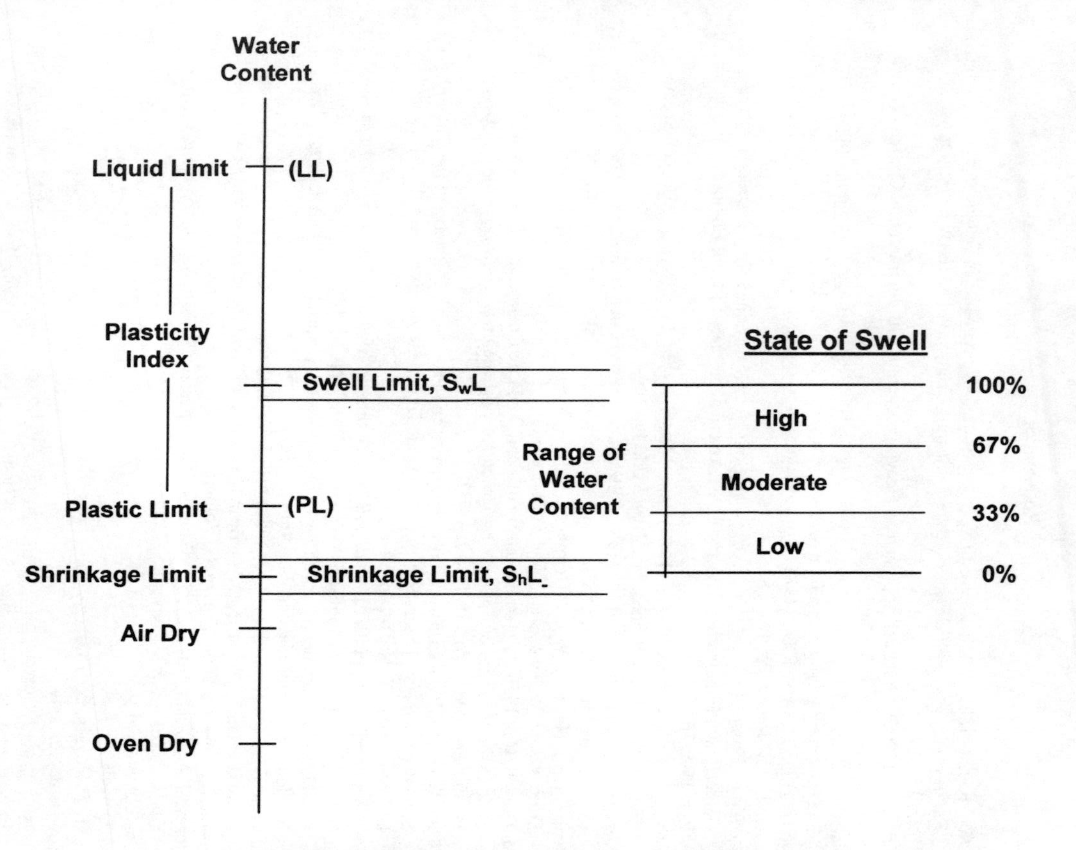

Fig. 1: Soil Moisture Scale with Atterberg Limits Related to State of Swell

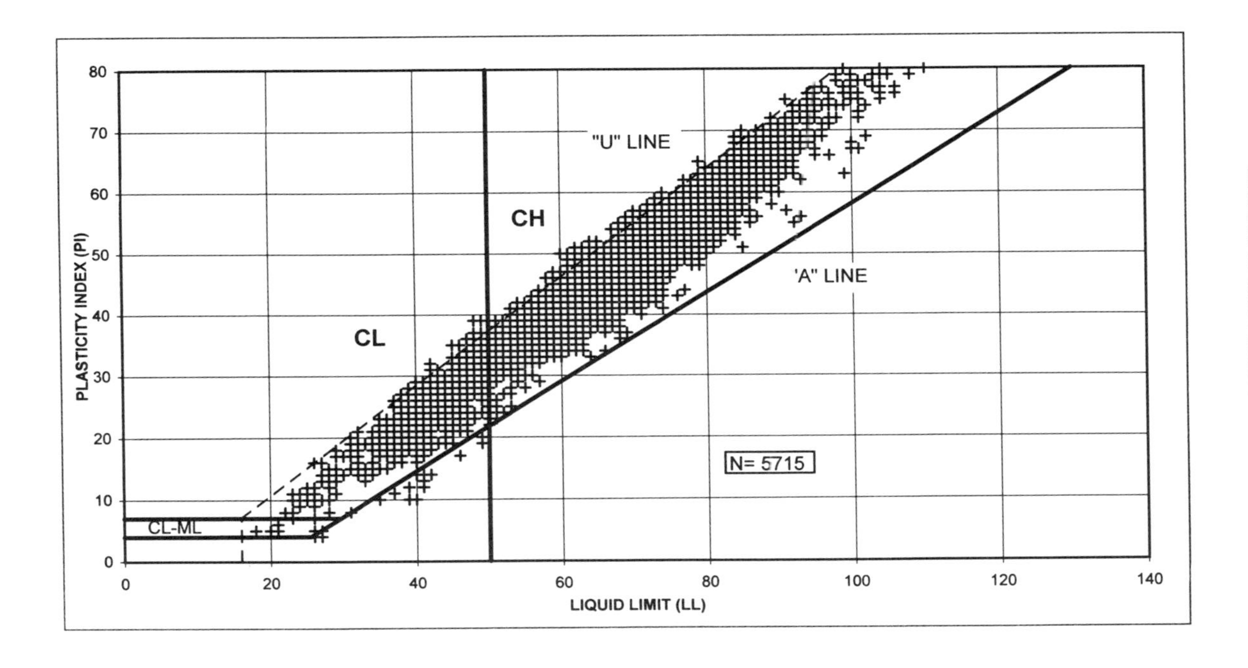

Figure 2: Casagrande Liquid Limit vs Plasticity Index: San Antonio Database

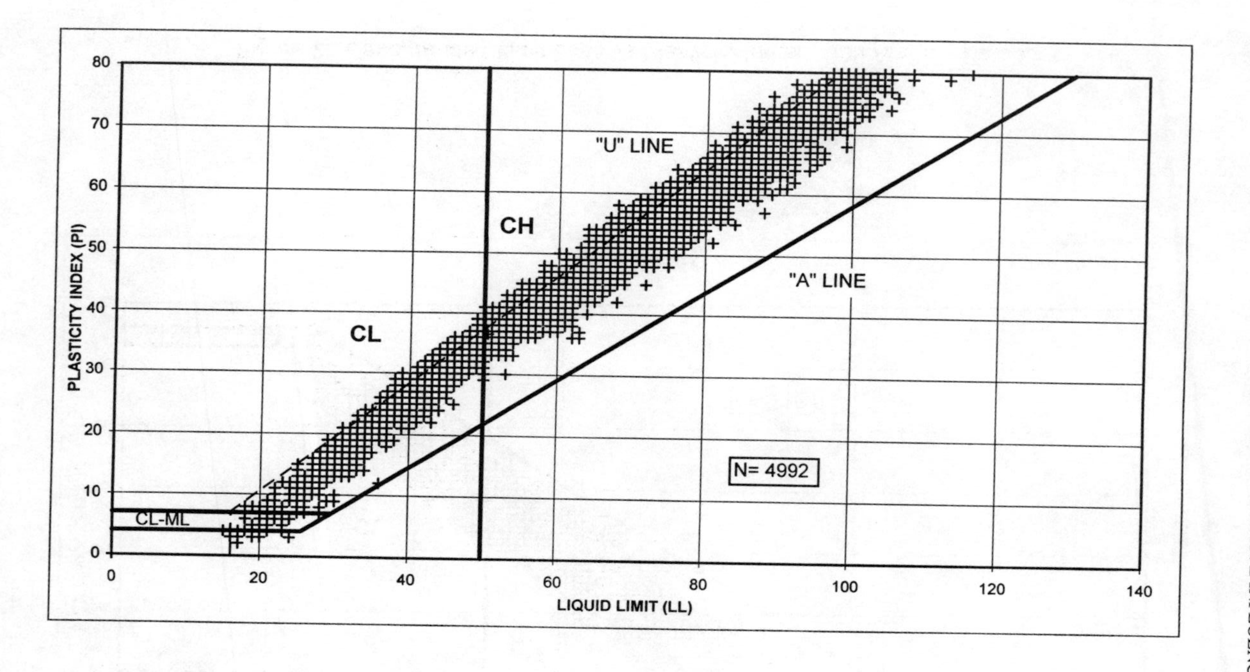

Figure 3: Casagrande Liquid Limit vs Plasticity Index: Houston Database

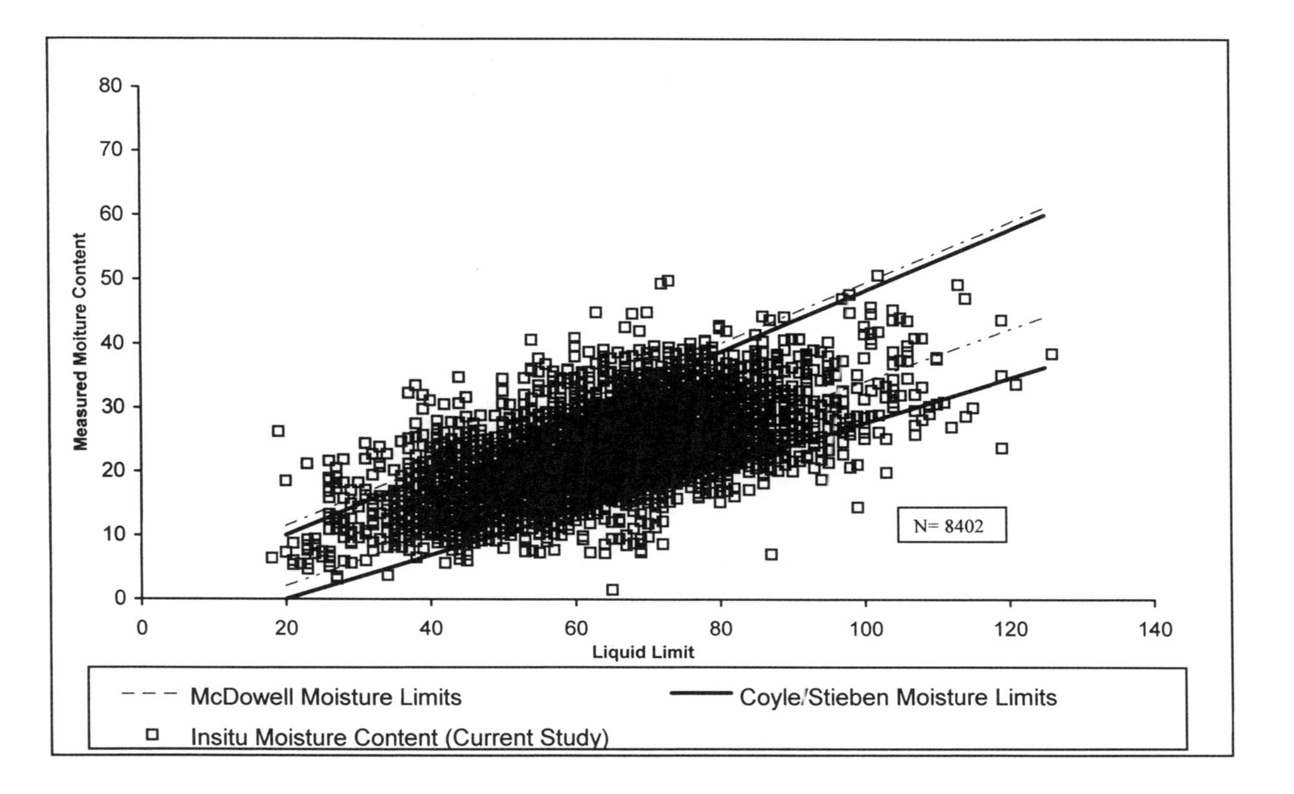

Figure 4: Comparison of Moisture Data from Current Study to McDowell and Coyle Moisture Limits: San Antonio Database

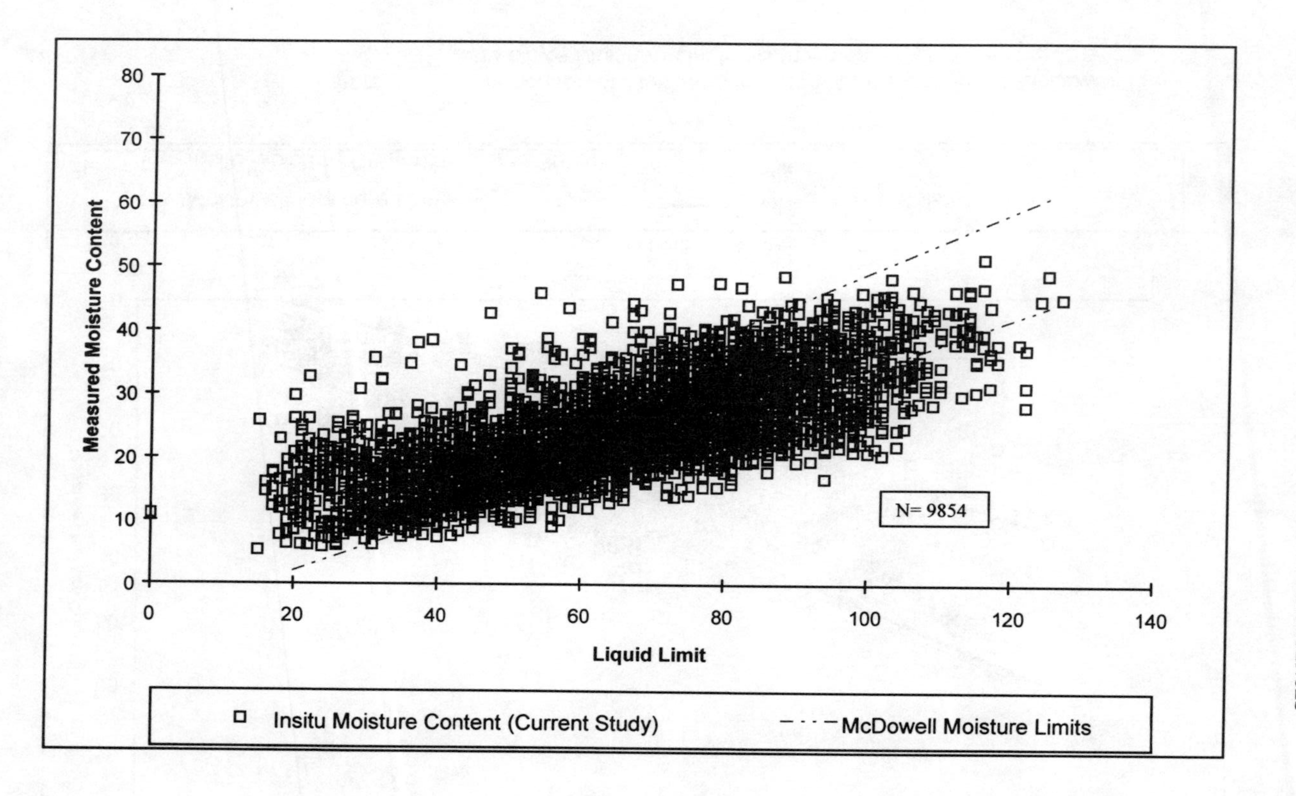

Figure 5: Comparison of Moisture Data from Current Study to McDowell Moisture Limits: Houston Database

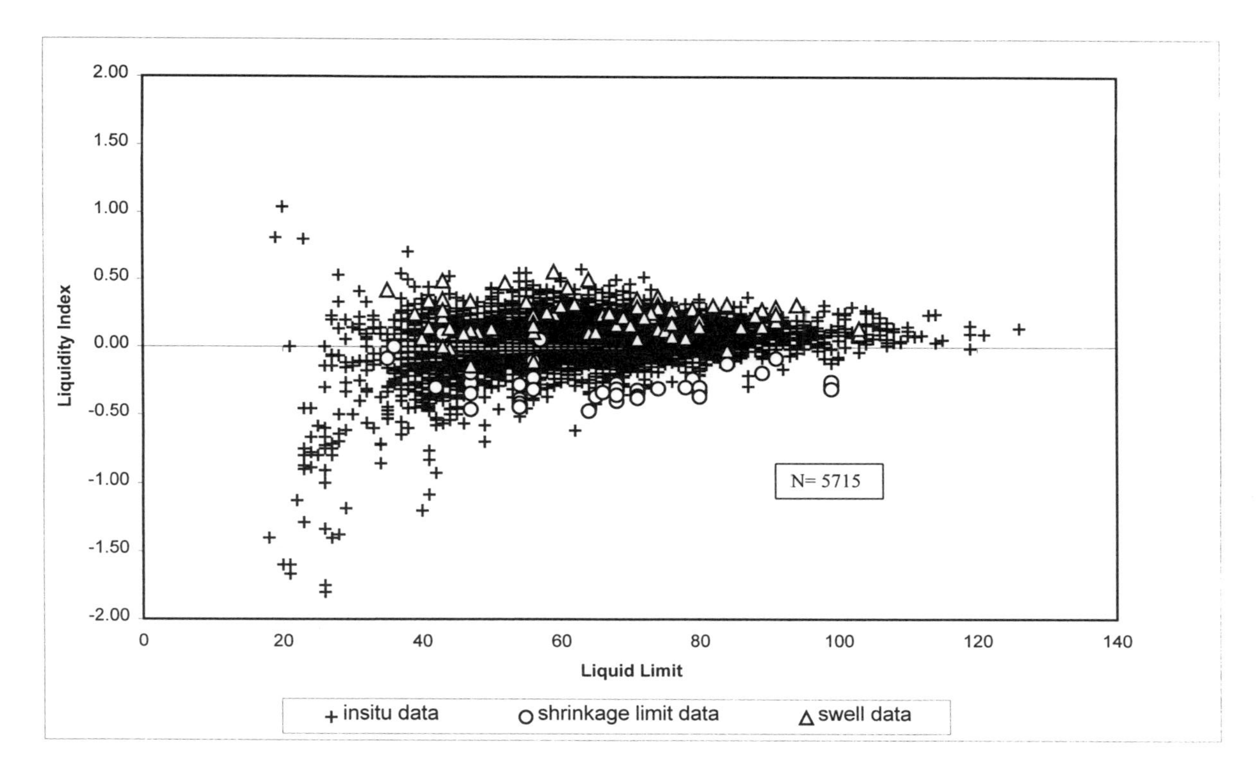

Figure 6:. Liquid Limit vs Liquidity Index – Insitu and Analytical Data: San Antonio Database

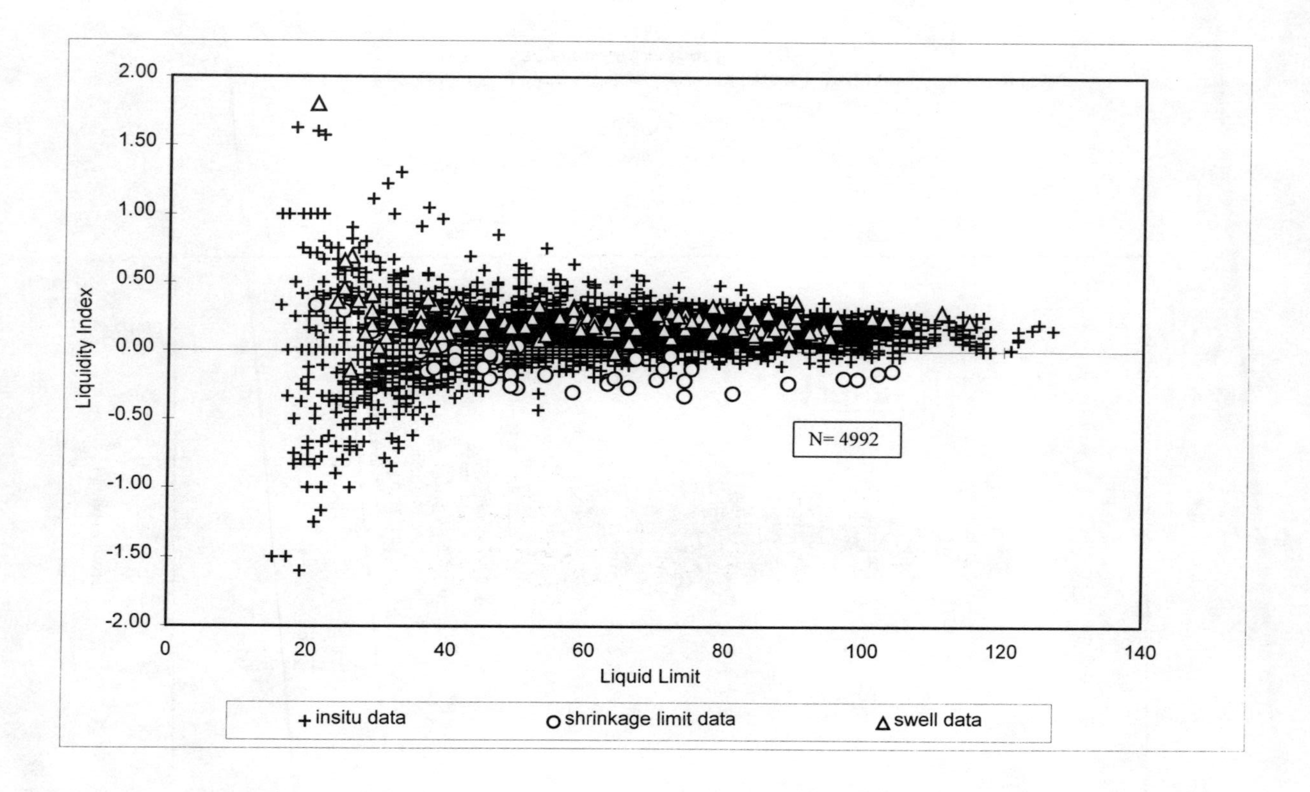

Figure 7:. Liquid Limit vs Liquidity Index – Insitu and Analytical Data: Houston Database

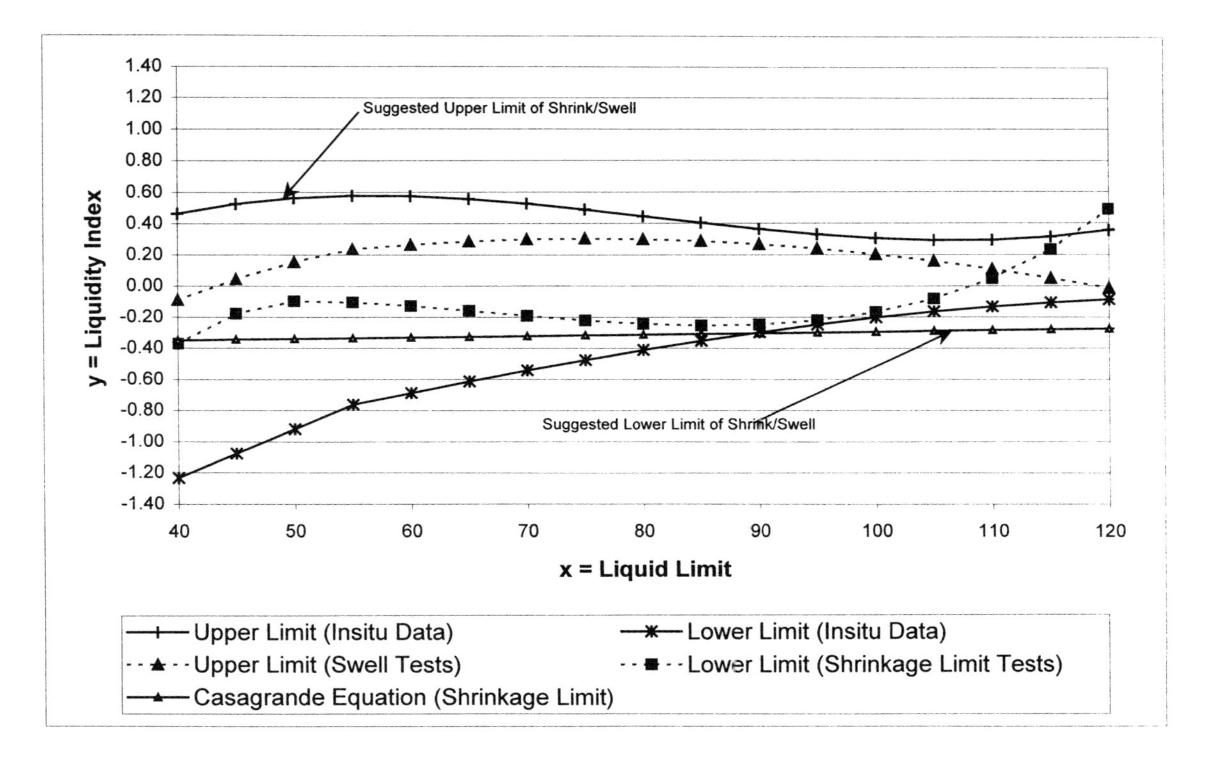

Figure 8: Approximated Regression Boundaries of Shrinkage vs Swell
For San Antonio (Pecan Gap)

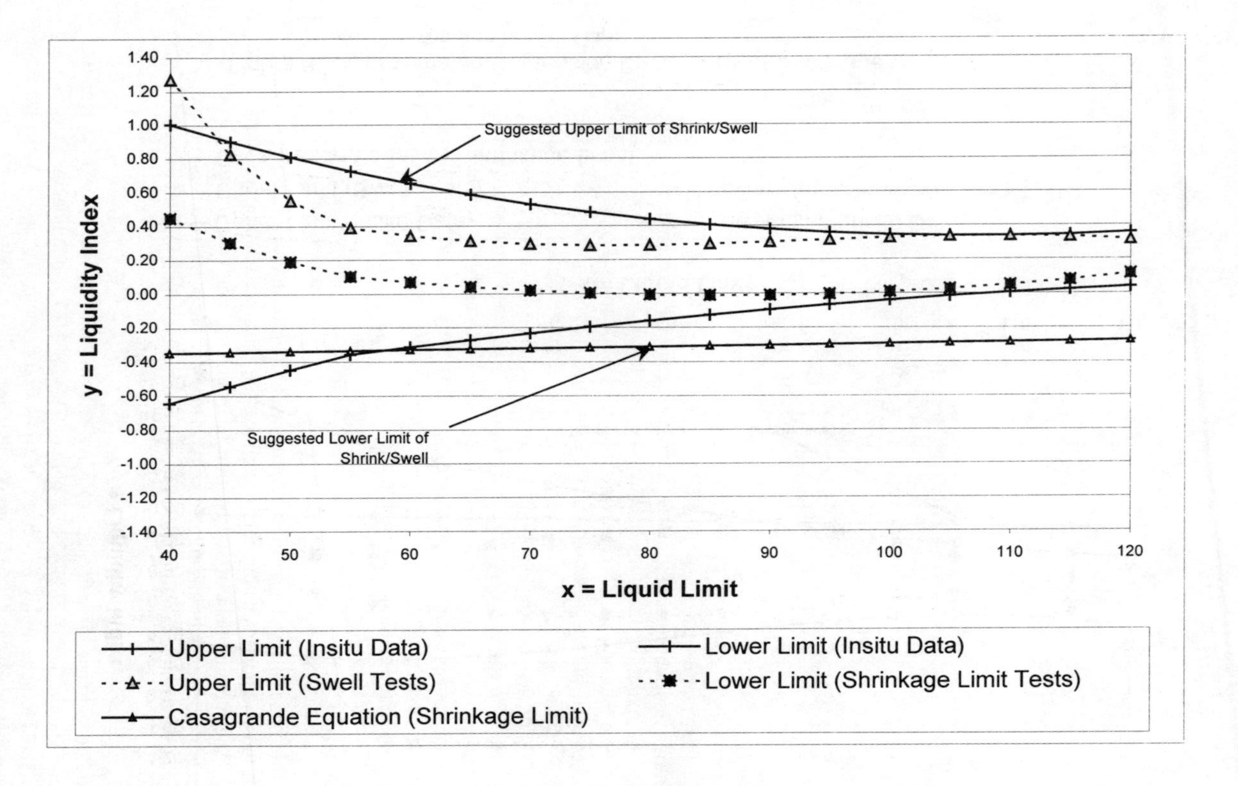

Figure 9: Approximated Regression Boundaries of Shrinkage vs Swell
For Houston (Beaumont)

Soil Suction Measurements by Filter Paper

Rifat Bulut[1], M.ASCE, Robert L. Lytton[2], F.ASCE, and Warren K. Wray[3], F.ASCE

Abstract

This paper reports on an evaluation of wetting and drying filter paper suction calibration and soil total and matric suction measurement techniques of filter paper method. Calibration of the method was investigated by constructing two calibration curves; one by using the process of wetting the filter papers through vapor flow and the other by using the method of drying the filter papers through fluid flow. The wetting curve was constructed using sodium chloride (NaCl) salt solutions and Schleicher & Schuell No. 589-WH filter papers. It was found that the change in the wetting suction curve is very sensitive to minor changes in filter paper water content below about 1.5 log kPa (2.5 pF) suction. The drying curve was established by employing both pressure plate and pressure membrane devices and the same filter papers. In developing the filter paper calibration curves, the capabilities, pitfalls, and limitations of the method are also discussed.

Introduction

The filter paper method is a soil suction measurement technique. Soil suction is one of the most important parameters describing the moisture condition of unsaturated soils. The measurement of soil suction is crucial for applying the theories of the engineering behavior of unsaturated soils. The filter paper method is an inexpensive and relatively simple laboratory test method, from which both total and matric

----

[1]Graduate Student, Department of Civil Engineering, Texas A&M University, College Station, Texas 77843-3136; phone 979-458-4147; r-bulut@tamu.edu.
[2]A.P. and Florence Wiley Professor of Civil Engineering, Texas A&M University, College Station, Texas 77843-3136; phone 979-845-8211; r-lytton@tamu.edu.
[3]Provost and Senior Vice President for Academic and Student Affairs, Michigan Technological University, Houghton, Michigan 49931-1295; wkwray@mtu.edu.

suction measurements are possible. With a reliable soil suction measurement technique, the initial and final soil suction profiles can be obtained from samples taken at convenient depth intervals. The change in suction with seasonal moisture movement is valuable information for many engineering applications.

This paper evaluates calibration techniques for filter paper wetting and drying processes, and soil total and matric suction measurements with filter paper method by construction of two calibration curves. The wetting curve was constructed using NaCl salt solutions and Schleicher & Schuell No. 589-WH filter papers. Salt solutions and filter papers were brought to equilibrium through vapor flow (filter paper wetting process) at isothermal conditions. Equilibrium time and temperature were two weeks and 25°C, respectively. The temperature was maintained at 25°C within ± 0.1°C fluctuations. The drying curve was established using both pressure plate and pressure membrane devices and the same filter papers. The pressure plate apparatus can measure matric suction values up to 150 kPa. However, with the pressure membrane device matric suction values can be extended up to 10,000 kPa. The equilibration periods were selected as 3, 5, and 7 days depending on the testing set up, which will be described below.

A Brief Historical Background

There are many soil suction measurement techniques and instruments in the fields of soil science and engineering. Most of these instruments have limitations with regard to range of measurement, equilibration times, and cost. Therefore, there is a need for a method which can cover the practical suction range, be adopted as a basis for routine testing, and is inexpensive. One of those soil suction measurement techniques is the filter paper method, which was evolved in Europe in the 1920s and came to the United States in 1937 with Gardner (1937). Since then, the filter paper method has been used and investigated by numerous researchers (Fawcett and Collis-George 1967; McQueen and Miller 1968; Al-Khafaf and Hanks 1974; McKeen 1980; Hamblin 1981; Chandler and Guierrez 1986; Houston et al. 1994; Swarbrick 1995), who have tackled different aspects of the filter paper method. Different types of materials were used, such as filter papers and suction measuring devices, and different experimental techniques to calibrate the filter paper and to measure suction of the soil sample. Therefore, it is very difficult to compare these methods on a one-to-one basis.

All the calibration curves established from Gardner (1937) to Swarbrick (1995) appear to have been constructed as a single curve by using different filter papers, a combination of different soil suction measuring devices, and different calibrating testing procedures. However, Houston et al. (1994) developed two different calibration curves; one for total suction and one for matric suction measurements using Fisher quantitative coarse filter papers. For the total suction calibration curve, saturated salt solutions and for the matric suction calibration curve tensiometers and pressure membranes were employed. Houston et al. (1994) reported that the total and matric suction calibration curves were not compatible. This simply implies that two different calibration curves, one for matric and one for total suction, need to be used in soil suction measurements. However, in this paper the fact is presented that the two curves reflect an expected hysteresis between wetting and drying effects and that the appropriate curve for both matric and total suction is the wetting curve since this matches the process that the filter paper undergoes in the measurement process.

Soil Suction Concept

In general, porous materials have a fundamental ability to attract and retain water. The existence of this fundamental property in soils is described in engineering terms as suction, negative stress in the pore water. In engineering practice, soil suction is composed of two components: matric and osmotic suction (Fredlund and Rahardjo 1993). The sum of matric and osmotic suction is called total suction. Matric suction comes from the capillarity, texture, and surface adsorptive forces of the soil. Osmotic suction arises from the dissolved salts contained in the soil water. This relationship can be formed in an equation as follows:

$$h_t = h_m + h_\pi \tag{1}$$

where h_t = total suction (kPa), h_m = matric suction (kPa), and h_π = osmotic suction (kPa).

Total suction can be calculated using Kelvin's equation, which is derived from the ideal gas law using the principles of thermodynamics and is given as:

$$h_t = \frac{RT}{V} \ln\left(\frac{P}{P_o}\right) \tag{2}$$

where h_t = total suction, R = universal gas constant, T = absolute temperature, V = molecular volume of water, P/P_o = relative humidity, P = partial pressure of pore water vapor, and P_o = saturation pressure of water vapor over a flat surface of pure water at the same temperature.

If Eq. (2) is evaluated at a reference temperature of 25°C, the following total suction and relative humidity relationship can be obtained:

$$h_t = 137182 \times \ln(P/P_o) \tag{3}$$

Figure 1 shows a plot of Eq. (3) at 25°C temperature. From Fig. 1, it can be seen that there is nearly a linear relationship between total suction (h_t) and relative humidity (P/P_o) over a very small relative humidity range. It can be said, in general, that in a closed system under isothermal conditions the relative humidity may be associated with the water content of the system such as 100 percent relative humidity refers to a fully saturated condition. Therefore, the suction value of a soil sample can be inferred from the relative humidity and suction relationship if the relative humidity is evaluated in some way. In a closed system, if the water is pure enough, the partial pressure of the water vapor at equilibrium is equal to the saturated vapor pressure at temperature, T. However, the partial pressure of the water vapor over a partly saturated soil will be less than the saturation vapor pressure of pure water due to the soil matrix structure and the free ions and salts contained in the soil water (Fredlund and Rahardjo 1993).

In engineering practice, soil suction has usually been calculated in *pF* units (Schofield 1935) (i.e., suction in $pF = log_{10}(|suction\ in\ cm\ of\ water|)$). However, soil suction is also currently being represented in *log kPa* unit system (Fredlund and Rahardjo 1993) (i.e., suction in $log\ kPa = log_{10}(|suction\ in\ kPa|)$). The relationship between these two systems of units is approximately *suction in log kPa = suction in pF – 1.*

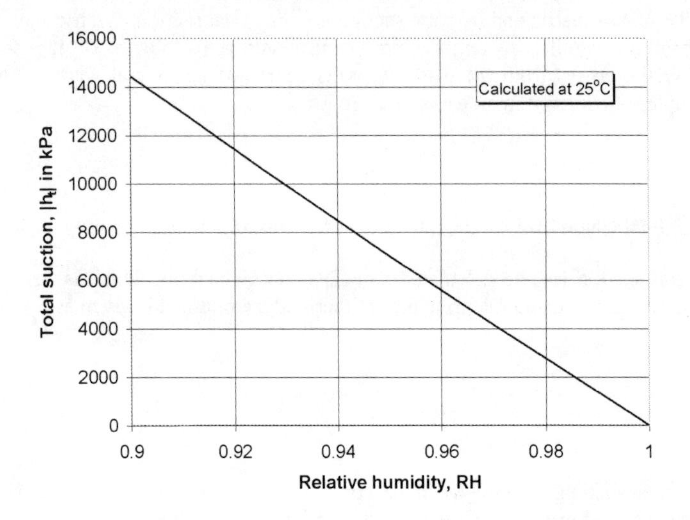

Figure 1. Total Suction versus Relative Humidity.

If total suction in kPa from Fig. 1 is converted to log kPa units, Fig. 2 is obtained. The difference between Fig. 1 and Fig. 2 is only the suction unit. The suction unit in Fig. 1 is kPa whereas it is log kPa in Fig. 2. From Fig. 2 it can clearly be seen that when relative humidity approaches 100 percent, the total suction becomes very sensitive. The sensitivity in the suction is due to the common logarithm used to convert suction from kPa to the log kPa unit.

Matric suction can be calculated from pressure plate and pressure membrane devices as the difference between the applied air pressure and water pressure across a porous plate. Matric suction can be formed in a relationship as follows:

$$h_m = -(u_a - u_w) \tag{4}$$

where h_m = matric suction, u_a = applied air pressure, and u_w = free water pressure at atmospheric condition.

The osmotic suction of electrolyte solutions, that are usually employed in the calibration of filter papers and psychrometers, can be calculated using the relationship between osmotic coefficients and osmotic suction. Osmotic coefficients are readily available in the literature for many different salt solutions. Table 1 gives the osmotic coefficients for several salt solutions. Osmotic coefficients can also be obtained from the following relationship (Lang 1967):

$$\phi = -\frac{\rho_w}{vmw} \ln\left(\frac{P}{P_o}\right) \tag{5}$$

where ϕ = osmotic coefficient, v = number of ions from one molecule of salt (i.e., v = 2 for NaCl, KCl, NH_4Cl and v = 3 for Na_2SO_4, $CaCl_2$, $Na_2S_2O_3$, etc.), m = molality, w = molecular mass of water, and ρ_w = density of water.

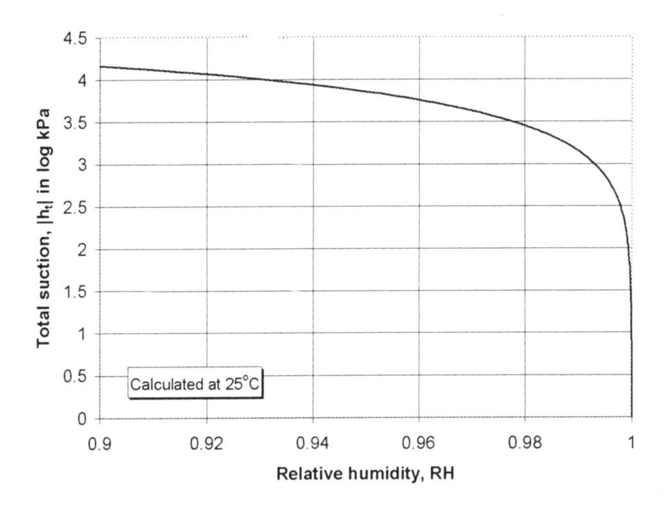

Figure 2. Total Suction and Relative Humidity Relationship.

The relative humidity term (P/P_o) in Eq. (5) is also known as the activity of water (a_w) in physical chemistry of electrolyte solutions. The combination of Eq. (2) and Eq. (5) gives a useful relationship that can be adopted to calculate osmotic suctions for different salt solutions:

$$h_\pi = -vRTm\phi \tag{6}$$

Table 2 gives osmotic suctions for several salt solutions using osmotic coefficients from Table 1 and Eq. (6).

Table 1. Osmotic Coefficients of Several Salt Solutions.

	Osmotic Coefficients at 25°C						
Molality (m)	$NaCl$[a]	KCl[a]	NH_4Cl[a]	Na_2SO_4[b]	$CaCl_2$[c]	$Na_2S_2O_3$[b]	$MgCl_2$[c]
0.001	0.9880	0.9880	0.9880	0.9608	0.9623	0.9613	0.9627
0.002	0.9840	0.9840	0.9840	0.9466	0.9493	0.9475	0.9501
0.005	0.9760	0.9760	0.9760	0.9212	0.9274	0.9231	0.9292
0.010	0.9680	0.9670	0.9670	0.8965	0.9076	0.8999	0.9106
0.020	0.9590	0.9570	0.9570	0.8672	0.8866	0.8729	0.8916
0.050	0.9440	0.9400	0.9410	0.8229	0.8619	0.8333	0.8708
0.100	0.9330	0.9270	0.9270	0.7869	0.8516	0.8025	0.8648
0.200	0.9240	0.9130	0.9130	0.7494	0.8568	0.7719	0.8760
0.300	0.9210	0.9060	0.9060	0.7262	0.8721	0.7540	0.8963
0.400	0.9200	0.9020	0.9020	0.7088	0.8915	0.7415	0.9206
0.500	0.9210	0.9000	0.9000	0.6945	0.9134	0.7320	0.9475
0.600	0.9230	0.8990	0.8980	0.6824	0.9370	0.7247	0.9765
0.700	0.9260	0.8980	0.8970	0.6720	0.9621	0.7192	1.0073
0.800	0.9290	0.8980	0.8970	0.6629	0.9884	0.7151	1.0398
0.900	0.9320	0.8980	0.8970	0.6550	1.0159	0.7123	1.0738
1.000	0.9360	0.8980	0.8970	0.6481	1.0444	0.7107	1.1092
1.200	0.9440	0.9000	0.8980	...	...	...	...
1.400	0.9530	0.9020	0.9000	...	...	...	...
1.500	...	...	...	0.6273	1.2004	0.7166	1.3047
1.600	0.9620	0.9050	0.9020	...	...	...	...
1.800	0.9730	0.9080	0.9050	...	...	...	...
2.000	0.9840	0.9120	0.9080	0.6257	1.3754	0.7410	1.5250
2.500	1.0130	0.9230	0.9170	0.6401	1.5660	0.7793	1.7629

References:
[a]Hamer and Wu, 1972
[b]Goldberg, 1981
[c]Goldberg and Nuttall, 1978

The Filter Paper Method

The filter paper method has long been used in soil science and engineering practice and it has recently been accepted as an adaptable test method for soil suction measurements because of its advantages over other suction measurement devices. Basically, the filter paper comes to equilibrium with the soil either through vapor (total suction measurement) or liquid (matric suction measurement) flow. At equilibrium, the suction value of the filter paper and the soil will be equal. After equilibrium is established between the filter paper and the soil, the water content of the filter paper disc is measured. Then, by using a filter paper water content versus suction calibration curve, the corresponding suction value is found from the curve.

This is the basic approach suggested by ASTM Standard Test Method for Measurement of Soil Potential (Suction) Using Filter Paper (ASTM D 5298). In other words, ASTM D 5298 employs a single calibration curve that has been used to infer both total and matric suction measurements. The ASTM D 5298 calibration curve is a combination of both wetting and drying curves. However, this paper demonstrates that the "wetting" and "drying" suction calibration curves do not match, an observation that was also made by Houston et al. (1994).

Table 2. Osmotic Suctions of Several Salt Solutions.

	Osmotic Suctions in kPa at 25°C						
Molality (m)	NaCl	KCl	NH_4Cl	Na_2SO_4	$CaCl_2$	$Na_2S_2O_3$	$MgCl_2$
0.001	5	5	5	7	7	7	7
0.002	10	10	10	14	14	14	14
0.005	24	24	24	34	34	34	35
0.010	48	48	48	67	67	67	68
0.020	95	95	95	129	132	130	133
0.050	234	233	233	306	320	310	324
0.100	463	460	460	585	633	597	643
0.200	916	905	905	1115	1274	1148	1303
0.300	1370	1348	1348	1620	1946	1682	2000
0.400	1824	1789	1789	2108	2652	2206	2739
0.500	2283	2231	2231	2582	3396	2722	3523
0.600	2746	2674	2671	3045	4181	3234	4357
0.700	3214	3116	3113	3498	5008	3744	5244
0.800	3685	3562	3558	3944	5880	4254	6186
0.900	4159	4007	4002	4384	6799	4767	7187
1.000	4641	4452	4447	4820	7767	5285	8249
1.200	5616	5354	5343	...	...	...	...
1.400	6615	6261	6247	...	...	...	...
1.500	...	...	...	6998	13391	7994	14554
1.600	7631	7179	7155	...	...	...	...
1.800	8683	8104	8076	...	...	...	...
2.000	9757	9043	9003	9306	20457	11021	22682
2.500	12556	11440	11366	11901	29115	14489	32776

Calibration for the Suction Wetting Curve

The calibration for the suction wetting curve for filter paper using salt solutions is based upon the thermodynamic relationship between total suction (or osmotic suction) and the relative humidity resulting from a specific concentration of a salt in distilled water. The thermodynamic relationship between total suction and relative humidity is given in Eq. (2).

In this study, NaCl was selected as an osmotic suction source for the filter paper calibration. Salt concentrations from 0 (distilled water) to 2.7 molality were prepared and filter papers were simply placed above salt solutions (in a non-contact manner) in sealed containers. The calibration test configuration adopted for this research is shown in Fig. 3. The filter paper and salt solution setups in the sealed containers were put in a constant temperature environment for equilibrium. Temperature fluctuations were kept as low as possible during a two week equilibration period. A water bath was employed for this purpose, in which temperature fluctuations did not exceed ± 0.1°C.

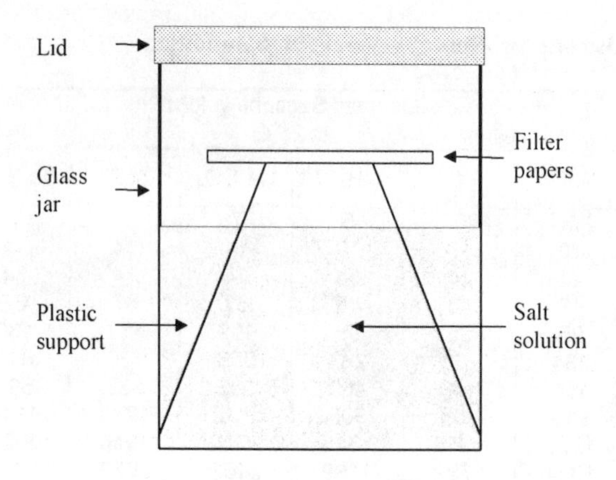

Figure 3. Total Suction Calibration Test Configuration.

Before commencing the filter paper calibration experiments and the soil suction measurements, all the items related to filter paper testing were cleaned carefully. Latex gloves and tweezers were used to handle the materials in nearly all steps of the experiment. The filter papers and aluminum cans for water content measurements were never touched with bare hands because oily hands may cause the filter papers to absorb more water. In addition, it is suggested that the filter paper water content measurements are performed by two persons in order to reduce the time during which the filter papers are exposed to the laboratory atmosphere and, thus, the amount of moisture lost or gained during measurements is kept to a minimum.

Experimental Procedure for Wetting Curve Calibration

The procedure that was adopted for the experiment is as follows:

1. NaCl solutions were prepared from 0 (i.e., distilled water) to 2.7 molality (i.e., the number of moles of NaCl in mass in 1,000 ml of distilled water).
2. A 250 ml glass jar was filled with approximately 150 ml of a solution of known molality of NaCl. Then, a small plastic cup was inserted into the glass jar to function as a support for filter papers. Two filter papers were put on the plastic cup one on top of the other. The glass jar lid was sealed tightly with plastic tapes to ensure air tightness. The configuration of the setup is depicted in Fig. 3.
3. Step 2 was repeated for each different NaCl concentration.

The glass jars were inserted into large plastic containers and the containers were sealed with water proof tape. Then, the containers were put into sealed plastic bags for extra protection. After that, the containers were inserted into the water bath for an equilibration period. After two weeks of equilibrating time, the procedure for the filter paper water content measurements was as follows:

1. Before taking the plastic containers from the water bath, all aluminum cans were weighed to the nearest 0.0001 g. accuracy and recorded on a filter paper water content measurement data sheet, similar to the one provided in ASTM D 5298.
2. After that, all measurements were carried out by two persons. For instance, while one person was opening the sealed glass jar, the other person was transferring the filter paper, using tweezers, into the aluminum can very quickly (i.e., in a few seconds, usually less than 5 seconds). The lid was placed on each aluminum can immediately.
3. Then, the weights of each can with filter papers inside were very quickly measured to the nearest 0.0001 g.
4. Steps 2 and 3 were followed for every glass jar. Then, all the cans were put into the oven with the lids half-open to allow evaporation. All filter papers were kept at 105 ± 5°C temperature for 24 hours inside the oven. This is the standard test method for soil water content measurements. However, it is only necessary to keep the filter paper in the oven for at least 10 hours.
5. Before taking measurements, the cans were closed with their lids and allowed to equilibrate in the oven for about 5 minutes. Then, a can was removed from the oven and put on an aluminum block for about 20 seconds to cool down; the aluminum block acted as a heat sink and expedited the cooling of the can. This is to eliminate temperature fluctuations and air currents in the enclosed weighing scale. After that, the can with dry filter paper inside was weighed to the nearest 0.0001 g. very quickly. The dry filter paper was taken out of the can and the cooled can was also weighed very quickly.
6. Step 5 was repeated for every can.

Wetting Calibration Curve

A wetting curve was constructed from the filter paper test results by following the procedure described above. The curve obtained for Schleicher & Schuell No. 589-WH filter papers using sodium chloride salt solutions is depicted in Fig. 4. Figure 4 clearly shows the sensitivity of total suction to very small changes in filter paper water content values when the relative humidity approaches 100%. The reason behind this sudden drop in suction was briefly explained with Fig. 2 and it will be discussed in detail in the following paragraphs.

There is an inverse relationship between total suction and relative humidity at a constant temperature (i.e., Eq. (2)). Figure 1 was obtained by plotting Kelvin's equation for 25°C temperature. From the relationship, total suction is equal to zero when relative humidity is 100 percent (i.e., fully saturated condition). On the other hand, total suction becomes very large when relative humidity decreases, but the change in relative humidity is very small with respect to the change in total suction. For instance, a relative humidity of 94 percent at a temperature of 25°C corresponds

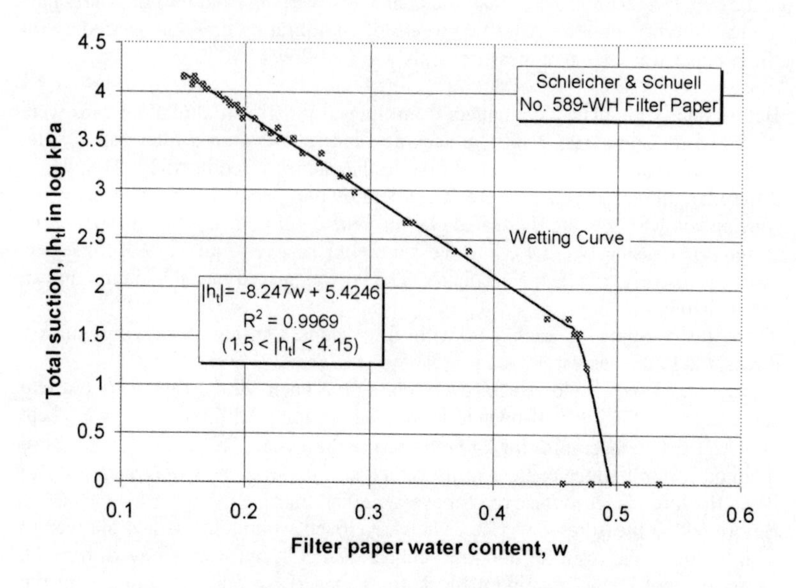

Figure 4. Filter Paper Wetting Calibration Curve.

to a total suction value of 8,488 kPa. Since the total suction values in engineering practice are often represented in logarithmic scales (i.e., pF or log kPa), the total suction values in log kPa units versus relative humidity were plotted in Fig. 2 in order to see the effect of the logarithmic scale on the relationship. From the figure, it is seen that total suction decreases dramatically when relative humidity approaches 100 percent.

Different concentrations of sodium chloride solutions were plotted against corresponding osmotic (or total) suction values both in kPa and log kPa units at 25°C in Figs. 5 and 6, respectively. As expected, the trend of the curves are similar to the trend of the curves obtained for relative humidity versus total suction for both kPa and log kPa units in Figs. 1 and 2, respectively. For example, a high concentration salt solution at a constant temperature in a closed container has low relative humidity above its surface.

Figure 7 depicts a plot of the wetting curve in kPa units versus filter paper water contents obtained in this research. In other words, if the suction values in Fig. 4 are plotted in kPa units, Fig. 7 is obtained. From the figure, the sensitivity of the filter paper water contents and total suction relationship can clearly be seen at very low suction values. From the relationships between total suction and relative humidity (i.e., Figs. 1 and 2), total suction and salt solutions (i.e., Figs. 5 and 6), and total suction and filter paper water contents (i.e., Figs. 4 and 7) it can be concluded that the dramatic decrease in total suction at high water contents depends on the nature of the relationship between total suction and relative humidity from Kelvin's equation and on the use of the logarithmic scale for total suction. In addition, soils

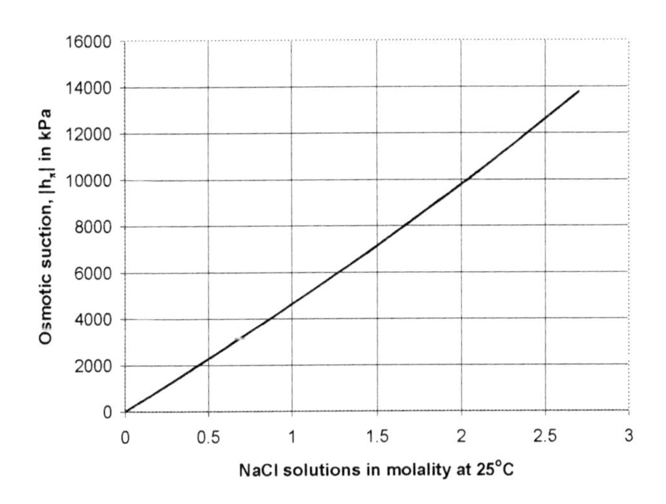

Figure 5. Osmotic Suction versus NaCl Solutions.

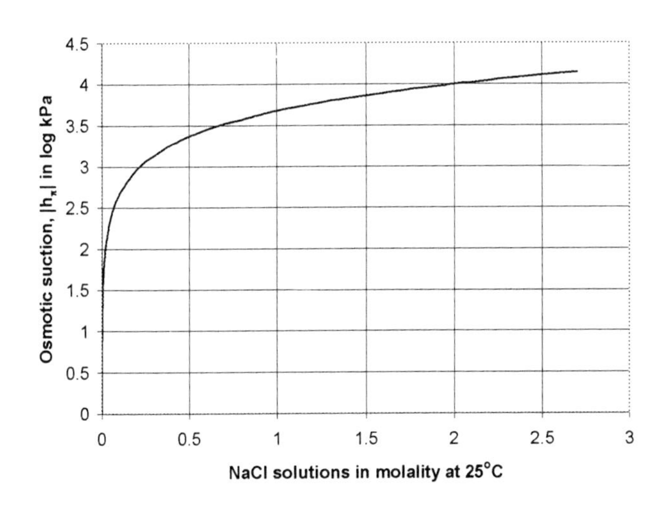

Figure 6. Osmotic Suction versus NaCl Solutions at 25°C.

tend to absorb more water for a small change in suction at very low suction values (Baver et al. 1972), and since filter papers, like soils, are porous materials they are very sensitive for absorbing water at low suction values.

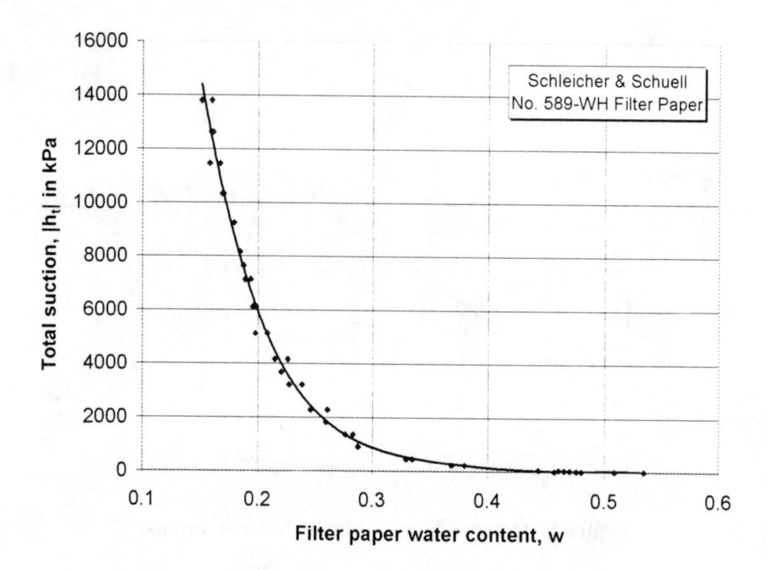

Figure 7. Filter Paper Wetting Calibration Curve in kPa Units.

Calibration for the Suction Drying Curve

Pressure plate and pressure membrane devices were employed in the drying filter paper calibration. A schematic drawing of a pressure plate or pressure membrane apparatus is depicted in Fig. 8. For the drying suction calibration of the filter paper, a contact path is provided between the filter paper and the measuring device so as to eliminate the osmotic suction component of total suction. In other words, if transfer of the soil water is allowed only through fluid flow, dissolved salts will move with the soil water, and the measuring device will not detect the osmotic suction component.

Pressure plate and pressure membrane devices operate by imposing a suction value (i.e., applied air pressure minus water pressure at atmospheric condition) on a given specimen which can be a soil or filter paper. The filter paper is put into the suction measuring device in a manner that ensures good contact with the porous plate or cellulose membrane. In this process, the main concern is to make sure that an intimate contact is provided between the water inside the filter paper and the water inside the porous disk so that transfer of the water is allowed only through continuous water films. To investigate the degree of contact between the filter paper and porous disk, the testing procedure and setup as depicted in Fig. 8 were undertaken in this study. Three different soils (i.e., a fine clay, sandy silt, and pure sand) were used in the calibration process of filter papers in order to investigate the role of soils in establishing a good contact between the filter paper and porous disk.

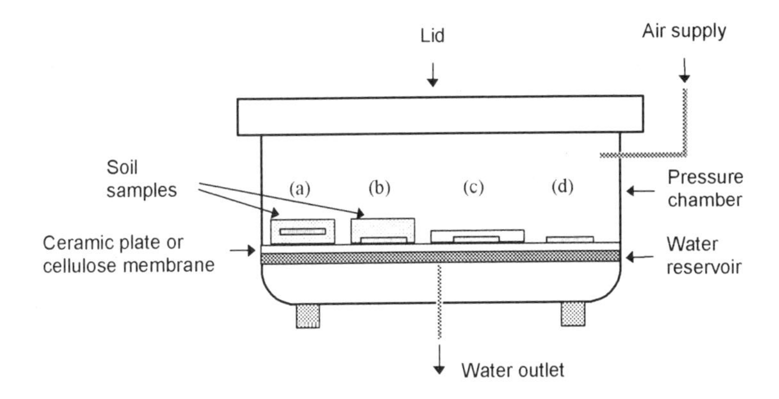

(a) One filter paper between two larger size protective filter papers embedded into the soil sample.
(b) One filter paper makes contact with the porous plate or membrane and covered on top with a larger size protective filter paper in the soil sample.
(c) One filter paper makes contact with the porous plate or membrane and covered on top with two larger size protective filter papers.
(d) One filter paper on the porous plate or membrane.

Figure 8. Schematic Drawing of a Pressure Plate or Membrane Device.

Experimental Procedure for Drying Curve Calibration

The procedure that was adopted for the experiment is as follows:

1. Prior to each test, the porous disk or membrane and the soils were saturated with distilled water at least one day in advance, so that all the pores were fully saturated with water.
2. The testing configuration as in Fig. 8 was established using one of the soils (i.e., fine clay or sandy silt or pure sand). Figure 8 explains how the filter papers, soil, and protective papers were arranged in the experiment. The soil specimens with the filter papers were placed on the saturated disks and the level of distilled water on the plate was raised enough to cover all of the filter papers. All of the air bubbles were eliminated during placement of the filter paper, soil, and protective paper arrangement on the ceramic disk by carefully pressing the bubbles out to the edges of each. The air bubbles were pressed out of the sample using a small diameter glass pipe and a large diameter glass cylinder.

3. After the pressure chamber was tightened, with the influence of the applied air pressure the water inside the soil specimen and filter papers were forced out through the porous plate or membrane and collected in a graduated cylinder until a suction equilibrium between the soil and filter papers and the applied air pressure was established.

An equilibration period between 3 and 5 days is commonly suggested for matric suction measurements using pressure plates and membranes (ASTM D 5298, Houston et al. 1994, Lee 1991). The equilibrating periods used for this study varied between 3, 5, and 7 days depending on the testing set up. For instance, when filter papers were embedded in the soil, equilibrating periods were 7 days for the fine clay and 5 days for the sandy silt set up, but the equilibrating period was 3 days when filter papers embedded in the pure sand or when only filter papers were used. However, all the three soils were also tested with filter papers inside in the same pressure chamber to check the differences between the filter paper water contents. To obtain the filter paper water contents, the same procedure described in the Wetting Curve Calibration Procedure was followed.

Drying Calibration Curve

A drying curve was established from the filter paper test results by following the procedure described above. The curve obtained for Schleicher & Schuell No. 589-WH filter papers using both pressure plate and pressure membrane devices is depicted in Fig. 9. Each data point on Fig. 9 is an average of at least three tests and each test data is an average of at least four filter papers. The standard errors for the straight line and curved portions of the drying curve are 0.135 and 0.116 log kPa units, respectively. The standard error for the straight line portion of the wetting curve is 0.044 log kPa. With the pressure membrane the highest matric suction obtained was 4,570 kPa and suctions below 150 kPa were obtained using the pressure plate apparatus. The corresponding wetting calibration curve is also shown in Fig. 9. It plots below the drying suction curve, as is expected of the hysteresis process.

Very high filter paper water contents were obtained when all the three soils were used as in the set up (a) as shown in Fig. 8. However, the filter paper water contents were all comparable as obtained from the set ups (b), (c), and (d) as in Fig. 8. The results from (b) were slightly wetter than (c) and the results from (d) were slightly drier than (c). In obtaining the calibration curve, the filter papers from the set up arrangements (b), (c), and (d) were used.

Soil Total and Matric Suction Measurements

Soil total suction measurements are similar to those measurements in the filter paper calibration testing. The same testing procedure can be followed by replacing the salt solution with a soil sample.

Soil matric suction measurements are also similar to the total suction measurements except that an intimate contact should be provided between the filter paper and the soil. A suggested testing procedure for soil total and matric suction measurements using filter papers is outlined in Appendix.

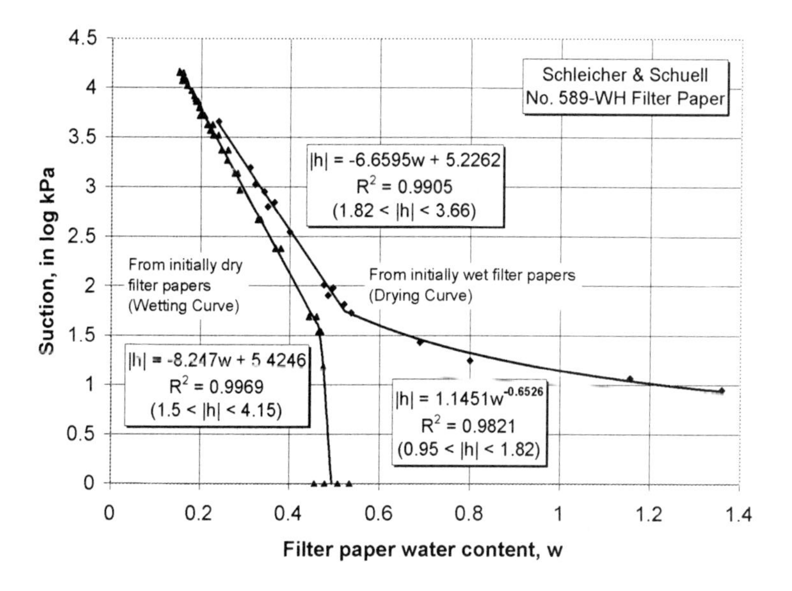

Figure 9. Drying Suction Calibration Curve along with Wetting Suction Curve.

Discussion

The dramatic decrease of total suction at high filter paper water contents is related to the nature of Kelvin's equation and to the use of the logarithmic scale (i.e., log kPa or pF). These results conclude that the filter paper method can give reliable wetting suction results up to a point. In other words, with the Schleicher & Schuell No. 589-WH filter papers reliable wetting suction measurements can be taken at and above 1.5 log kPa (2.5 pF), but below about 1.5 log kPa wetting suction results cannot be relied upon because a small error in measuring water content can result in a large error in the inferred suction. Therefore, a best fit line up to 1.5 log kPa was made to plot Fig. 4, below which there is a sudden drop in the wetting suction.

In the drying filter paper calibration testing, filter papers are initially fully saturated and with the application of air pressure the water inside the filter paper is driven out, which is a drying process. However, the soil matric and total suction measurements follow a wetting process with the filter paper method. Because of hysteresis, the wetting suction calibration curve must always plot below the drying calibration curve. A final point; because both the matric and total suction measurements are wetting processes, they should, by these arguments, both be determined from the wetting calibration curve.

Acknowledgements

The authors wish to thank Dr. Hsiu-Chung Lee for helping with the total suction calibration testing and Dr. Seong-Wan Park for helping with the matric suction calibration testing.

References

Al-Khafaf, S. and Hanks, R. J. (1974). "Evaluation of the Filter Paper Method for Estimating Soil Water Potential," *Soil Science*, Vol. 117, No. 4, pp. 194-199.

Baver, L. D., Gardner, W. H., and Gardner, W. R. (1972). *Soil Physics*, John Wiley & Sons, Inc., New York.

Chandler, R. J. and Gutierrz, C. I. (1986). "The Filter Paper Method of Suction Measurements," *Geotechnique*, Vol. 36, pp. 265-268.

Fawcett, R. G. and Collis-George, N. (1967). "A Filter-Paper Method for Determining the Moisture Characteristics of Soil," *Australian Journal of Experimental Agriculture and Animal Husbandry*, Vol. 7, pp. 162-167.

Fredlund, D. G. and Rahardjo, H. (1993). *Soil Mechanics for Unsaturated Soils*, New York: John Wiley & Sons, Inc.

Gardner, R. (1937). "A Method of Measuring the Capillary Tension of Soil Moisture Over a Wide Moisture Range," *Soil Science*, Vol. 43, No. 4, pp. 277-283.

Goldberg, R. N. (1981). "Evaluated Activity and Osmotic Coefficients for Aqueous Solutions: Thirty-six Uni-Bivalent Electrolytes," *Journal of Physics and Chemistry Reference Data*, Vol. 10, No. 3, pp. 671-764.

Goldberg, R. N. and Nuttall, R. L. (1978). "Evaluated Activity and Osmotic Coefficients for Aqueous Solutions: The Alkaline Earth Metal Halides," *Journal of Physics and Chemistry Reference Data*, Vol. 7, No. 1, pp. 263-310.

Hamblin, A. P. (1981). "Filter Paper Method for Routine Measurement of Field Water Potential," *Journal of Hydrology*, Vol. 53, No. 3/4, pp.355-360.

Hamer, W. J. and Wu, Y.-C. (1972). "Osmotic Coefficients and Mean Activity Coefficients of Uni-Univalent Electrolytes in Water at 25°C," *Journal of Physics and Chemistry Reference Data*, Vol. 1, No. 4, pp. 1047-1099.

Houston, S. L., Houston, W. N., and Wagner, A. M. (1994). "Laboratory Filter Paper Measurements," *Geotechnical Testing Journal*, GTJODJ, Vol. 17, No. 2, pp. 185-194.

Lang, A. R. G. (1967). "Osmotic Coefficients and Water Potentials of Sodium Chloride Solutions from 0 to 40°C," *Australian Journal of Chemistry*, Vol. 20, pp. 2017-2023.

Lee, H. C. (1991). "An Evaluation of Instruments to Measure Soil Moisture Condition," *M.Sc. Thesis,* Texas Tech University, Lubbock, Texas.

McKeen, R. G. (1980). "Field Studies of Airport Pavement on Expansive Clay," *Proceedings 4th International Conference on Expansive Soils,* Vol. 1, pp.242-261, ASCE, Denver, Colorado.

McQueen, I. S. and Miller, R. F. (1968). "Calibration and Evaluation of a Wide-Range Gravimetric Method for Measuring Moisture Stress," *Soil Science*, Vol. 106, No. 3, pp. 225-231.

Schofield, R. K. (1935). "The pF of the Water in Soil," *Transactions, 3rd International Congress of Soil Science*, Vol. 2, pp. 37-48.
Swarbrick, G. E. (1995). "Measurement of Soil Suction Using the Filter Paper Method," *First International Conference on Unsaturated Soils*, Eds.: E. E. Alonso and P. Delage, Vol. 2, Paris, 6-8 September, ENDPC, pp. 701-708.

Appendix. Soil Suction Measurements

Soil Total Suction Measurements

Glass jars that are between 250 to 500 ml volume size are readily available in the market and can be easily adopted for suction measurements. Glass jars, especially, with 3.5 to 4 inch (8.89 to 10.16 cm) diameter can contain the 3 inch (7.62 cm) diameter Shelby tube samples very nicely. A testing procedure for total suction measurements using filter papers can be outlined as follows:

Experimental Procedure

1. At least 75 percent by volume of a glass jar is filled up with the soil; the smaller the empty space remaining in the glass jar, the smaller the time period that the filter paper and the soil system requires to come to equilibrium.
2. A ring type support, which has a diameter smaller than filter paper diameter and about 1 to 2 cm in height, is put on top of the soil to provide a non-contact system between the filter paper and the soil. Care must be taken when selecting the support material; materials that can corrode should be avoided, plastic or glass type materials are much better for this job.
3. Two filter papers one on top of the other are inserted on the ring using tweezers. The filter papers should not touch the soil, the inside wall of the jar, and underneath the lid in any way.
4. Then, the glass jar lid is sealed very tightly with plastic tape.
5. Steps 1, 2, 3, and 4 are repeated for every soil sample.
6. After that, the glass jars are put into the ice-chests in a controlled temperature room for equilibrium.

Researchers suggest a minimum equilibrating period of one week (ASTM D 5298, Houston et al. 1994, Lee 1991). After the equilibration time, the procedure for the filter paper water content measurements can be as follows:

1. Before removing the glass jar containers from the temperature room, all aluminum cans that are used for moisture content measurements are weighed to the nearest 0.0001 g. accuracy and recorded.
2. After that, all measurements are carried out by two persons. For example, while one person is opening the sealed glass jar, the other is putting the filter paper into the aluminum can very quickly (i.e., in a few seconds) using tweezers.
3. Then, the weights of each can with wet filter paper inside are taken very quickly.

4. Steps 2 and 3 are followed for every glass jar. Then, all cans are put into the oven with the lids half-open to allow evaporation. All filter papers are kept at 105 ± 5°C temperature inside the oven for at least 10 hours.
5. Before taking measurements on the dried filter papers, the cans are closed with their lids and allowed to equilibrate for about 5 minutes. Then, a can is removed from the oven and put on an aluminum block (i.e., heat sinker) for about 20 seconds to cool down; the aluminum block functions as a heat sink and expedites the cooling of the can. After that, the can with the dry filter paper inside is weighed very quickly. The dry filter paper is taken from the can and the cooled can is weighed again in a few seconds.
6. Step 5 is repeated for every can.

After obtaining all of the filter paper water contents an appropriate calibration curve, such as the one in Fig. 4, is employed to get total suction values of the soil samples.

Soil Matric Suction Measurements

Soil matric suction measurements are similar to the total suction measurements except instead of inserting filter papers in a non-contact manner with the soil for total suction testing, a good intimate contact should be provided between the filter paper and the soil for matric suction measurements. Both matric and total suction measurements can be performed on the same soil sample in a glass jar as shown in Fig. A1. A testing procedure for matric suction measurements using filter papers can be outlined as follows:

Experimental Procedure

1. A filter paper is sandwiched between two larger size protective filter papers. The filter papers used in suction measurements are 5.5 cm in diameter, so either a filter paper is cut to a smaller diameter and sandwiched between two 5.5 cm papers or bigger diameter (bigger than 5.5 cm) filter papers are used as protectives.
2. Then, these sandwiched filter papers are inserted into the soil sample in a very good contact manner (i.e., as in Fig. A1). An intimate contact between the filter paper and the soil is very important.
3. After that, the soil sample with embedded filter papers is put into the glass jar container. The glass container is sealed up very tightly with plastic tape.
4. Steps 1, 2, and 3 are repeated for every soil sample.
5. The prepared containers are put into ice-chests in a controlled temperature room for equilibrium.

Researchers suggest an equilibration period of 3 to 5 days for matric suction testing (ASTM D 5298, Houston et al. 1994, Lee 1991). However, if both matric and total suction measurements are performed on the same sample in the glass jar, then the final equilibrating time will be at least 7 days of total suction equilibrating period. The procedure for the filter paper water content measurements at the end of the equilibration is exactly same as the one outlined for the total suction water content measurements. After obtaining all the filter paper water contents the appropriate calibration curve may be employed to get the matric suction values of the soil samples.

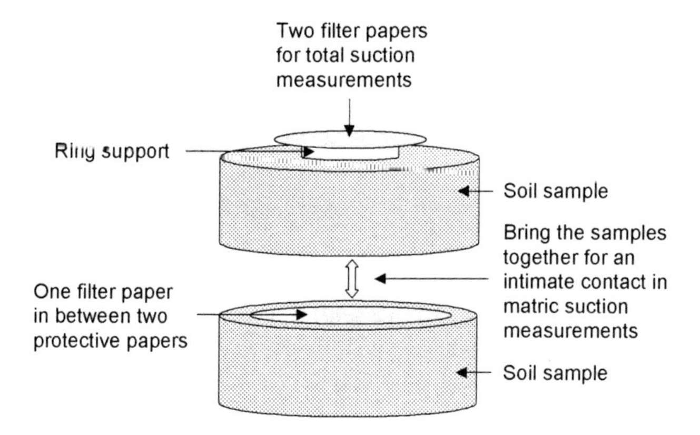

Figure. A1. Total and Matric Suction Measurements.

Subject Index

Page number refers to the first page of paper

Author Index

Page number refers to the first page of paper